Livestock Sector Development of India and Potential Threats in WTO Regime

Livestock Sector Development of India and Potential Threats in WTO Regime

Deepak Shah

ELIVA PRESS

Published by Eliva Press
Email: info@elivapress.com
Website: www.elivapress.com

ISBN: 978-1-63648-326-9

© Eliva Press, 2021
© Deepak Shah
Cover Design: Eliva Press
Cover Image: Freepik Premium
Printed at: see last page

Livestock Sector Development of India and Potential Threats in WTO Regime

Deepak Shah

Professor,
Gokhale Institute of Politics and Economics (Deemed to be University),
Deccan Gymkhana, Pune 411004 (Maharashtra), India

1

Preface

Although dairy industry is the single largest contributor to India's GDP and involves over 80 million small farming households with its profound social impact, the opening up of the Indian market to an influx of foreign goods has raised much concern about the status of the Indian dairy industry in the era of WTO regime. The subsidies provided by the developed countries to their dairy farmers has helped them to lower their price of dairy products and consequently influencing the world prices, affecting, in turn, the farming community in developing world as this provides good opportunities for the traders to import cheaper milk products and thereby earn high profits at the expense of farmers belonging to developing world like India.

It is to be further noted that the WTO framework is based on free trade. However, the European Union along with United States have by passed and openly violated their commitments to WTO. In the dairy sector, the subsidised exports of EU have adversely affected the dairy industry in India, Brazil and Jamaica. In India, milk is produced by small farmers belonging to remote areas and processed in plants owned by cooperatives, whereas in EU countries and New Zealand there has been different concept where there stand factory style operations of milk production and squeezing their cows poses a great threat of dumping excess production at lower rates in rest of the world. Since under WTO provisions, there has been greater emphasis on liberalizing trade and government policies to enhance world import demand for dairy products, commitments on market access, reduction of domestic support and subsidies on exports for the removal of distortions in the domestic market, the consequent effect or pressure will be on small farmers, particularly on women. It is in the light of above background that this study attempts to assess not only the development of livestock sector over time in India coupled with its significance in the national economy and social welfare with emphasis on dairy sector but also examines the likely impact of trade liberalisation under WTO regime on the domestic market in general and the livelihood of farming community of India in particular. The paper pins attention to several aspects that need to be taken cognizance of to save Indian dairy industry from the great trade robbery indulged in by European countries in the era of WTO regime in which the cattle in the rich countries are pampered at the cost of several hundred million farmers in the developing world.

Deepak Shah

Contents

List of Tables

Livestock Sector Development of India and Potential Threats in WTO Regime

Deepak Shah[*]

Executive Summary

Despite constraints like rearing livestock under sub-optimal conditions due to low economic status of livestock owners, India has now become the largest producer of milk in the world due to dependence of millions of farmers on this secondary remunerative source of agricultural income. Development of Indian dairy sector is an unprecedented success story, which owes it to significant participation of women in this activity. Dairy industry is the single largest contributor to India's GDP and involves over 80 million small farming households with its profound social impact. However, the opening up of the Indian market to an influx of foreign goods has raised much concern about the status of the Indian dairy industry in the era of WTO regime. The subsidies provided by the developed countries to their dairy farmers has helped them to lower their prices of dairy products and consequently influencing the world prices and affecting the farming community in developing world as this provides good opportunities for traders to import cheaper milk products and earn high profits at the expense of farmers of developing world like India. In the present milieu, the Indian dairy sector would be competitive only if the export subsidies on dairy products are abolished. In more relaxed market environment, the real challenge before Indian livestock sector would be in terms of Sanitary and Phytosanitary Measures (SPS), Agreement on Technical Barriers to Trade (TBT) and animal welfare related issues. With a view to meet these requirements - both domestically and in the world markets - modernization of supply chain encompassing producer and consumer is the need of the hour.

Keywords: India, Livestock Development, WTO, Price Trends, Trade Distortions

---

[*] Professor, Gokhale Institute of Politics and Economics (Deemed to be University), Deccan Gymkhana, Pune 411004 (Maharashtra), India [Email: deepak.ds.shah@gmail.com]

4

Livestock Sector Development of India and Potential Threats in WTO Regime

Deepak Shah[*]

Backdrop

For ages, man has been husbanding livestock with the aim of improving their quality and making them more useful. Animal husbandry in India has been closely integrated with agriculture and plays a pivotal role in ensuring socio-economic development of rural masses. Among various animal husbandry activities in India, dairying is treated as the most important one since milk produced through bovines not only serves as one of the main sources of proteins and calcium for a largely vegetarian population but its consumption by far has become essential for human health.

In India, agriculture is an economic symbiosis of crop and milk production. Historically, the role of livestock has been complementary to crop production. Dairying and agriculture are bound together by a set of mutual input-output relationships. Dairying is not an adjunct to the crop-mix of Indian farms but an integral part of the total farming system. Hence, treating dairy cattle as the backbone of the livestock wealth of our country would not be an exaggeration. Although dairying provides livelihood to millions of Indian farmers and generates additional income and employment for a large number of families in the countryside with significantly greater participation of women in this important activity, our country with about 18.36 per cent of the world's total cattle and buffalo population accounts for only about 14.5 per cent of the world's total milk production (GOI, 2004 a). Our livestock are roughly half as efficient as the average milch animals in the world and only one-fifth as efficient as those in the advanced countries (Shah, 2001). There is consistent rise in India's share in world milk production, which has grown from 9.9 per cent to 14.5 per cent between 1990 and 2003 (Appendix 1).

The milk grid in this country is based on the average produce of millions of small-uninformed farmers who are still unaware of modern scientific ways of dairy farming. This is perceived as a major handicap, as the industry is unable to attain the quality requirements to compete in the world market. Therefore, technologies should be adopted to make milk production system viable and sustainable to usher in an era of quality consciousness to compete internationally. The moot point to consider over here is

---

[*] Professor, Gokhale Institute of Politics and Economics (Deemed to be University), Deccan Gymkhana, Pune 411004 (Maharashtra), India [Email: deepak.ds.shah@gmail.com]

whether this is achievable with the kind of animals and resources poor dairy farmers have at their disposal. Undoubtedly, despite several limiting factors, dairy industry of India has undergone considerable transformation in due course of time mainly due to the application of scientific production techniques by medium and large farmers and greater importance being given to the development of dairy cooperative infrastructure that has contributed in no small measure towards substantial growth in milk production since the early seventies. The investment, effort, innovation and energy of our farmers and industry have seen India moving from insignificant to becoming a major player in the world dairy scene. It is heartening to note that India today ranks as the world's largest milk producer and the value of output through dairying is the largest as compared to any other agricultural commodity.[1] The gross value of output from livestock sector at current prices has grown more than ten folds from Rs.10,599 crores in 1980-81 to Rs.1,39,981 crores in 2000-01 and further to Rs.1,64,509 crores in 2003-04 (CSO, 2005).

Although milk production in India has shown a rising trend ever since the inception of 'Operation Flood (OF)' programme in 1970-71, the Indian dairy industry acquired substantial growth from 8th Plan onwards with rise in milk production from 58 million tonnes in 1992-93 to 88.1 million tonnes in 2003-04. This has not only placed Indian dairy industry on the top of the world but also led to sustained growth in the availability of milk and milk products for the burgeoning population of the country. Dairying of late is considered as the secondary source of income for millions of rural households. The credit for this goes to OF (Box 1).

Box 1
Features of Operation Flood

In view of several positive features in favour of co-operative sector and practical results shown by milk co-operatives in Gujarat, it was decided by the government of India to extend institutional support in order to industrialise and organise all the dairy efforts in entire rural India through cooperatives. In 1965, the National Dairy Development Board (NDDB) was set up in India and in 1970-71 it drew up an all-encompassing programme known as 'Operation Flood'. The first phase of 'OF' programme began in July 1, 1970 and ended in March 31, 1981 with an investment of Rs.116.50 crores, covering 39 milk-sheds, 13,270 dairy cooperative societies (DCSs), 17.5 lakh members, 2.56 million kgs average daily milk procurement with 27.8 lakh litres per day liquid milk marketing. The second phase known as 'OF II' covered the period from April1, 1981 to March 31, 1985, with an investment of Rs.277.20 crores, covering 136 milk-sheds, 34,523 DCSs, 36.3 lakh members, 5.78 million kgs average daily milk procurement with 50.0 lakh litres per day liquid milk marketing. The 'OF III' programme began in April 1, 1987 and concluded on April 30, 1996 with an investment of Rs.137.95 crores, covering 170 milk-sheds, 72,744 DCSs, 93.0 lakh members, 11.0 million kgs average daily milk procurement with 110.0 lakh litres per day liquid milk marketing. During the first phase, only the states of Andhra Pradesh, Bihar, Delhi, Gujarat, Haryana, Karnataka, Madhya Pradesh, Maharashtra, Punjab, Rajasthan, Uttar Pradesh and West Bengal were covered under the programme. However, from second phase onwards almost all the states and majority of the Union Territories of India were covered under the programme.

At present, the efforts of the government is to promote dairy related activities in non-operation flood areas with emphasis on building up efficient cooperative infrastructure, rejuvenation of sick dairy cooperative federations and creation of infrastructure in the States for production of good quality milk and milk products[2] (GOI, 2005). In this context, an Integrated Dairy Development Programme in Non-Operation Flood, Hilly and Backward areas was launched during the 8th Plan, which continued in the 9th as well as in 10th Plan with a total outlay of Rs.175 crores as a Centrally Sponsored Plan Scheme with 100 per cent grants in aid basis to the States.[3] Since the inception of the scheme, 53 projects with the total outlay of Rs.292.19 crores have been sanctioned covering 149 districts in 23 States and one U.T. (GOI, 2005).

The National Agricultural Policy (NAP) document released in July 2000 relies heavily on 6-8 per cent growth rate in Animal Husbandry sector to achieve the targeted growth rate of 4 per cent for the Agriculture sector as a whole since the present rate of growth in crop production is around 2 per cent (GOI, 2004 b). The 10th Plan Approach Paper also stresses on the significance and importance of food and nutritional security through diversification of agriculture in animal husbandry and fishery sector. With a view to meet commitment to enhance food production in the country and achieve all round development of animal husbandry sector, government calls for rapid increase in the production of livestock products. The critical areas covered and strategies framed encompassing livestock sector under 10th Five Year Plan are shown in Box 2.

Box 2
Major Thrust Areas and Strategies for Livestock Sector under 10th Plan
The major thrust during the 10th Five Year Plan is on the following areas: ➢ Animal diseases control ➢ Livestock breed improvement and development ➢ Fodder development ➢ Dairy and Poultry development The strategies pursued for development of animal husbandry sector are as follows: ➢ Expansion of infrastructure and creation of seed stock of superior bulls and bull mothers ➢ Adequate animal health services ➢ Facilitate genetic improvement and conservation of indigenous breeds ➢ Improve productivity of pastureland through improved fodder seeds.

Though India produced around 80 million tonnes of milk in 2000-01, the organised sector in our country not even handled 10 per cent of this total milk production (NDDB, 2001). The consumer prices of milk in India are comparable to some of the lowest in the world due mainly to unremunerative and unattractive price offered to our dairy farmers for their milk production. Our purchasing power, and the demand for milk are not able to expand in line with the increasing milk production (Shah, 1997). In terms of per capita consumption of milk, India still compares poorly among the comity of world nations. An average per capita availability of only 231 gm milk per day during 2003-04 pushes our country at much lower place in the world in terms of milk availability. However, if our milk production continues to grow as it does now, we would have newer opportunities for launching a more meaningful marketing campaign. Though India has to do a lot of catching up to make her significant presence in world markets of milk and milk products, the withdrawal of Government regulations[4] from many spheres of economic and business activities has now enabled the use of cooperatives as an institutional set up for implementing the programmes relating to socio-economic development. The delicensing of dairy industry under Industrial development and Regulation Act (IDRA, 1951) and promulgation of Milk and Milk Products Order (MMPO)[5] in June 1992 have helped many milk unions in the country to increase their liquid milk collection and business turnover tremendously (Box 3).

Box 3
Milk and Milk Product Order, 1992

The Government of India had promulgated the Milk and Milk Product Order (MMPO) 1992 on 9.06.1992 under the provisions of Essential commodity Act, 1955 consequent to de-licensing of the dairy sector in 1991. As per the provisions of this order, any person/dairy plant handling more than 10000 liters per day of milk or 500 MT of milk solids per annum needs to be registered with the registering authority appointed by the Central Government. The main objective of the order is to maintain and increase in supply of liquid milk of desired quality in the interests of the general public and also for regulating the production, processing and distribution of milk and milk products.

Recognizing the necessity for suitable amendments in MMPO for faster pace of growth in the dairy sector, GOI has amended MMPO 1992 from time to time in order to make it more liberal and oriented to facilitate the dairy entrepreneurs. The GOI has notified the amendment proposals in the official gazette on 26/03/2002. Now, there is no restriction on setting up of new capacity, while noting that the requirement of registration is for enforcing the prescribed standards of quality and food safety.

The salient features of the new amendments made are as follows:

- The provision of assigning milk-shed has been done away with.
- The registration under MMPO-92 will now cover sanitary, hygienic condition, quality and food safety.
- The provision of inspection of dairy plant has been made flexible.
- The provision to grant registration in 90 days has been reduced to 45 days.
- The power of registration of State registering Authority has been raised from 1.00 LLPD to 2.00 LLPD.
- Altogether the Central and the State Registering Authorities have registered 688 units with combined capacity of 803.74 LLPD in Cooperative, Private and Government Sector as on 31.3.2003.

Though the withdrawal of Government regulations has helped many states, especially in terms of expansion and development of their dairy, it has been argued by Sharma et. al. (2003) that termination of licensing requirements for setting up milk processing and product manufacturing plants under the MMPO in 2002 led India's dairy one of the most deregulated industries in the world. In dismal contrast, prior to OF, when the link between the producer and consumer was missing and exports of almost all dairy products had been banned, the processing activity was controlled through licensing that favoured cooperatives over private entrepreneurs. It is believed by Sharma et. al. (2003) that the kind of deregulation allowed under MMPO coupled with a rapid increase in domestic and global demand for milk and milk products and distortions in the world dairy markets will virtually lead to several social and economic problems.

Ironically, though dairy industry is the single largest contributor to India's GDP and involves over 80 million small farming households with its profound social impact, the opening up of the Indian market to an influx of foreign goods has raised much concern about the status of the Indian dairy industry in the era of WTO regime. The subsidies provided by the developed countries to their dairy farmers has helped them to lower their price of dairy products and consequently influencing the world prices, affecting, in turn, the farming community in developing world as this provides good opportunities for the traders to import cheaper milk products and thereby earn high profits at the expense of farmers belonging to developing world like India.

Despite the fact that WTO framework is based on free trade, European Union along with United States by passed and openly violated their commitments to WTO. For instance, New Zealand Dairy Board dumped a large quantity of butter oil into India at price below $1000 per ton, though the prevailing international price is around $ 1300 per ton. New Zealand's butter oil was made available at Rs.64.54 per kg that stood cheaper by Rs.15 a kg as compared to prevailing international prices of Rs.87.40 per kg. Consequently, domestic prices of butter oil crashed by 10 15 per cent as the recent import was in the range of Rs.100 to Rs.120 per kg. Though consumer gained from this situation, the Indian producers were the worst affected and faced music of cheap imports.

It is in the light of above background that this study attempts to assess not only the development of livestock sector in India coupled with its significance in the national economy and social welfare with emphasis on dairy sector but also examines the likely impact of trade liberalisation under WTO regime on the domestic market in general and the livelihood of farming community of India in particular. It pins attention to several aspects that need to be taken cognizance of to save Indian dairy industry from the great trade robbery indulged in by European countries in the era of WTO regime in which the cattle in rich countries are pampered at the cost of several hundred million farmers in the developing world. The free world trade regime ushered in by the WTO not only poses threats to India's dairy industry but also opens up many opportunities for the industry. This is certainly a point that needs to be investigated in the today's WTO regime.

Dairy Development in India

India's economic efficiency as a milk producer is essentially a product of the Indian milk production system. It is a subsidiary activity in the rural economy closely intertwined with crop-agriculture and low cost feed and labour. At the farm level, technical efficiency as measured by yield remains low (Shah, 2001). There is, therefore, ample scope for further improvement in the efficiency of Indian dairying to make it globally more competitive. There is also every indication that the rate of growth of dairy industry is likely to gather further momentum due to cost-push effect and the consequent higher returns to the milk producers. Nonetheless, before going into the details of such issues it is essential to delve into the development of this sector in the country.

India's dairying scene has witnessed certain major changes in the last 25-30 years. Government outlay on development of livestock sector rose dramatically from a mere 905 million rupees in the Third Plan (1961-66) to the Sixth Plan (1980-85) total outlay of 3,966 million rupees on animal husbandry and dairying, of which 2,983 million rupees was meant for expenditure on dairying alone. During the Seventh Plan (1985-90), 3,028 million rupees was earmarked for dairying out of a total outlay of 4,679 million rupees for animal husbandry and dairy. Expenditure on dairying increased sharply during the Eighth Plan (1992-97). Of the total outlay of 13,000 million rupees for animal husbandry and dairying, the expenditure on dairying was nearly 63 per cent (Table 1). Nonetheless, though outlay on development of livestock sector increased to 15,456 million rupees in the Ninth Plan (1997-2002), only 30 per cent of the total outlay was earmarked for dairying and the remaining for the development of various other animal husbandry activities. The outlay for dairying in the Ninth Plan was substantially lower than the outlay for dairying in Eighth Plan but higher than the outlay for dairying in the Seventh Plan. Such increased allocation in plan outlay, leaving Ninth Plan aside, is a reflection of the importance of dairying in government's overall policy encompassing country's agricultural economy. Since dairying has already turned into viable and well developed sector, efforts of the government are now fully geared to strengthen other activities of livestock sector.

Table 1: Outlay and Expenditure of Central and Centrally Sponsored Schemes under Animal Husbandry and Dairying in India

(Rs. crores)

Plan	Period	Total Plan Outlay	Animal Husbandry		Dairying		Total		Expd. on Animal Husbandry to Total Outlay (%)	Expd. on Dairying to Total Outlay (%)
			Outlay	Expd.	Outlay	Expd.	Outlay	Expd.		
First	1951-56	1960.0	14.19	8.22	7.81	7.78	22.0	16.00	37.36	35.36
Second	1956-61	4600.0	38.50	21.42	17.44	12.05	55.94	33.47	38.29	21.54
Third	1961-66	8576.5	54.44	43.40	36.08	33.60	90.52	77.00	47.95	37.12
Annual	1966-69	6625.4	41.33	34.00	26.14	25.70	67.47	59.70	50.39	38.09
Fourth	1969-74	15778.8	94.10	75.51	139.00	78.75	233.10	154.26	32.39	33.78
Fifth	1974-78	39426.2	-	178.43	-	-	437.54	232.46	40.78	-
Sixth	1980-85	97500.0	60.46	39.08	336.10	298.34	396.56	337.42	9.85	75.23
Seventh	1985-90	180000.0	165.19	102.35	302.75	374.43	467.94	476.78	21.87	80.02
Annual	1990-91	-	43.71	36.18	79.67	41.43	123.38	77.61	29.32	33.58
Annual	1991-92	-	57.97	43.28	97.49	77.99	155.46	121.27	27.84	50.17
Eighth	1992-97	434100.1	400.00	305.43	900.00	818.05	1300.0	1123.5	23.49	62.93
	1992-93	80771.0	56.54	43.85	99.76	136.69	156.30	180.54	28.06	87.45
	1993-94	100120.1	78.26	54.59	257.74	216.44	336.00	271.03	16.25	64.42
	1994-95	112197.1	98.28	60.64	224.43	185.09	322.71	245.73	18.79	57.35
	1995-96	128590.0	94.00	66.66	250.00	179.67	344.00	246.33	19.38	52.23
	1996-97	-	103.94	81.04	155.98	100.29	259.92	181.33	31.18	38.58
Ninth	1997-2002	859200.0	1076.12	-	469.52	-	1545.6	-	-	-
	1997-98	-	160.15	94.84	39.00	29.24	199.15	124.08	47.62	14.68
	1998-99	-	170.40	53.03	50.60	23.97	221.00	77.00	24.00	10.85
	1999-2000	-	160.08	97.26	73.90	16.45	233.98	113.71	41.57	7.03
	2000-01	-	124.90	85.10	51.00	39.59	175.90	124.69	48.38	22.51
	2001-02	-	156.49	115.61	37.45	37.60	193.94	153.21	59.61	19.39

Source: Basic Animal Husbandry Statistics, 2002, Department of Animal Husbandry and Dairying, Ministry of Agriculture, Government of India, New Delhi

As a result of concerted efforts towards total dairy development, dairy industry in India has moved from dependence to self-reliance with total annual output of milk touching nearly 90 million tonnes of late that is valued at much higher than any other agro-based commodity produced in India. It is to be further noted that at present among 12 groups of food consumption items in the country, milk and milk products rank 3rd, next only to fruits and vegetables and cereals, both at current prices and at 1993-94 prices (Table 2). Though Table 2 also shows a decline in share of cereals, fruits and vegetables and milk and milk products in total private final consumption expenditure in domestic market between 1993-94 and 2003-04, this decline in share is only marginal in the case of milk and milk products (from 8 per cent to 7 per cent) as well as for fruits and vegetables (from 11 per cent to 10 per cent) as against cereals (from 14 per cent to 8 per cent), indicating not much effect on consumer expenditure on milk and milk products.

In fact, the National Commission on Agriculture (1972) in their Interim Report on milk production also recognized the importance of dairy sector and recommended that benefits of increasing demand for milk in large cities, towns and industrial area should go to small and marginal farmers and landless labourers. In India, landless labourers account for 21 per cent of total rural households. So they do not have any share in the total landholding. Nonetheless, they own 12 per cent of the milch animals and provide 16 per cent of all rural-produced milk. It stands to reason that dairying is a paying proposition for these poor rural people (Bedi, 1987). Hence efforts should be made to promote as much milk production as possible involving this segment of rural population. The commission suggested an integrated rural development approach based on a system of 'Kaira District Co-operative Milk Producers' Union Limited' commonly known as 'AMUL' in Anand of Gujarat (Jain, 1979).

Table 2: Private Final Consumption Expenditure By Object in India

(Rs. crores)

Sr. No.	Item	At Current Prices			At 1993-94 Prices		
		1993-94	2000-01	2003-04	1993-94	2000-01	2003-04
1.	Food, beverages & tobacco	315243	643300	778072	315243	387447	435865
		(54.85)	(47.30)	(44.06)	(54.85)	(47.27)	(45.17)
	Share in Private Consumption (%)						
	1.1 Food	50.60	42.23	38.26	50.60	42.48	39.33
	1.1.1 Cereals & bread	13.97	10.06	8.20	13.97	9.30	8.71
	1.1.2 Pulses	2.09	1.26	1.18	2.09	1.15	1.21
	1.1.3 Sugar & gur	3.51	2.93	2.02	3.51	3.39	2.75
	1.1.4 Oils & oilseeds	4.04	2.02	2.46	4.04	3.14	2.89
	1.1.5 Fruits & vegetables	10.89	9.97	9.73	10.89	9.72	8.47
	1.1.6 Potatoes & other tubers	1.08	0.77	0.67	1.08	0.95	0.80
	1.1.7 Milk & milk products	8.11	7.74	6.93	8.11	7.87	7.21
	1.1.8 Meat, egg & fish	3.78	3.75	3.51	3.78	3.61	3.78
	1.1.9 Coffee, tea & tobacco	1.02	0.66	0.47	1.02	0.72	0.66
	1.1.10 Spices	1.39	2.25	2.28	1.39	1.84	2.04
	1.1.11 Other food	0.74	0.81	0.79	0.74	0.79	0.82
	1.2 Beverages, pan & intoxicants	1.03	1.40	1.59	1.03	1.31	1.87
	1.2.1 Beverages	0.51	0.87	1.19	0.51	0.81	1.37
	1.2.2 Pan & other intoxicants	0.52	0.83	0.40	0.52	0.49	0.50
	1.3 Tobacco & its products	2.14	2.26	2.74	2.14	1.98	2.32
	1.4 Hotels & restaurants	1.07	1.40	1.47	1.07	1.50	1.65
2.	Clothing & footwear	34999	62609	77764	34999	43035	46037
3.	Gross rent, fuel & power	68239	155285	202049	68239	88674	97237
4.	Furniture, appliances, service, etc.	17610	40987	51256	17610	30123	34804
5.	Medical care & health services	19543	99338	146374	19543	41213	56596
6.	Transport & communication	64993	192876	274420	64993	122910	165427
7.	Recreation, education & cult. service	17626	48429	59549	17626	31100	37207
8.	Miscellaneous goods & services	36519	117194	176365	36519	75135	91692
	Private Final Consumption Expenditure in Domestic Market	574772	1360018	1765849	574772	819637	964865

Source: Computations are based on figures Compiled from National Accounts Statistics, Central Statistical Organisation, Ministry of Planning and Programme Implementation, Government of India, 2005

12

The success of the Kaira Union gave birth to other milk producer Unions in Gujarat. These milk producer Unions subsequently inspired the formation of 'National Dairy Development Board (NDDB)' in 1965 and provided all the impetus and resources required for its creation. Further, in view of several positive features in favour of milk co-operatives in Gujarat, it was finally decided by the government of India to extend institutional support in order to industrialize and organize all the dairy efforts in entire rural India, through co-operatives. In 1970-71 the NDDB drew up an all-encompassing programme known as 'Operation Flood', to replicate the Anand Pattern[6] Dairy Co-operatives in 18 areas of the country. The major objective of the operation flood programme was to build a viable and self-sustaining national dairy industry on co-operative lines. Total system approach was adopted for dairy development, which encompassed production, procurement, processing and marketing of milk. The efforts made by government towards development of this important sector of agriculture has paid rich dividends so far in terms of generating adequate employment and income to milk producers, and also in terms of value addition to the national economy.

Value Added by Livestock

In the current context of liberalization and increasing global integration of economies, it would be unfair on Indian dairy industry to compare it with that obtaining in most of the vastly modern and technologically far advanced western bloc countries in terms of a produce that is globally competitive. It should be realized that in India, the dairy industry is dependent on millions of small farmers who produce only a litre or two of marketable surplus and, it is this multitude of teeming million who eventually contribute to the overall flood of milk. The situation is entirely different in modern milk states. In most western countries, for instance, each of thousands of dairy farmers produces tonnes of milk. The sheer size and volume of production indulged in by milk producers in these countries makes it amenable for them to adopt meaningful scientific and technological means towards improving both quality and productivity (Shah, 2001). In contrast, the milk produced by our individual dairy farmers is so minimal that it is often very difficult to change their attitude in favour of modern animal husbandry practices that will make their produce cost effective as well as remunerative. An abysmally low production volume handled by our dairy farmers means introduction of any amount of technological innovation will not appreciably improve their income. Thus, mere technological innovation is not likely to transform the subsistence level dairy farmer into a market savvy commercial milk producer (Shah, 2001). Only real economic incentives and inducements can coerce and compel such farmers to change in favour of more profitable scientific farming. However, despite several weaknesses in terms of adoption of improved technique, the share of livestock in gross output of agriculture and allied activities has been showing a growing trend due mainly to dependence of millions of farmers on this secondary remunerative source of agricultural income.

The development of animal husbandry and dairying over time has resulted in significant expansion in value of livestock products in relation to value of agricultural products produced in India. The estimates relating to value of agricultural products and livestock products at different points of time encompassing the period between 1970-71 and 2003-04 are provided in Table 3.

Table 3: Value Added from Agriculture and Allied Activities in India: Current Prices

(Rs. crores)

Sr. No.	Items	At current prices				
		1970-71	1980-81	1990-91	2000-01	2003-04
1.	Value of Output	20730	56875	170698	518693	635104
	1.1 Agriculture	17531	46278	128657	378712	470595
	1.2 Livestock	3199	10597	42041	139981	164509
	Share (%)					
	a. Milk Groups	67.75	64.96	65.43	67.50	66.92
	b. Meat Groups	9.20	12.25	15.48	15.43	15.99
	c. Hides & Skins	2.29	2.36	1.66	1.84	1.83
	d. Eggs	6.63	3.31	3.11	3.17	2.93
	e. Wool & hair	0.61	0.46	0.34	0.23	0.18
	f. Dung	9.99	12.76	10.25	8.38	7.97
	g. Other Products	1.60	1.49	2.06	1.09	1.30
	i. Increment in Stock	1.93	2.41	1.67	2.36	2.88
2.	Less Inputs	4089	15247	38971	107020	127365
3.	Gross Domestic Product	16778	42466	135162	423523	521538
4.	Less Consumption of Fixed Capital	424	2418	7903	22083	27293
5.	Net Domestic Product	16354	40056	127259	401440	494245
6	Share in Value of Output					
	- Agriculture	84.57	81.37	75.37	73.01	74.10
	- Livestock	15.43	18.63	24.63	26.99	25.90
7.	Percent increase of 1.1 over					
	- 1970-71	-	163.98	633.88	2060.24	2584.36
	- 1980-81	-	-	178.01	718.34	916.89
	- 1990-91	-	-	-	194.36	265.77
	- 2000-01	-	-	-	-	24.26
8.	Percent increase of 1.2 over					
	- 1970-71	-	231.26	1214.19	4375.77	5042.51
	- 1980-81	-	-	296.73	1220.95	1452.41
	- 1990-91	-	-	-	232.96	291.31
	- 2000-01	-	-	-	-	17.52

Source: Estimates are based on figures Compiled from National Accounts Statistics, Central Statistical Organisation, Ministry of Planning and Programme Implementation, Government of India, 1982, 1996 & 2005

At current prices, the value of livestock products produced in the country in 2003-04 was estimated at Rs.1,64,509 crores with milk and milk products accounting for 67 per cent share in this value. Though the share of milk and milk products in total value added by livestock sector is significant and remained by and large constant over the past three decades, the growth in the value of milk product groups in absolute terms has increased by leaps and bounds as against other livestock based products produced in India during this period (Figure 1). It is to be further noted that the value of livestock products produced in the country has been steadily growing over the past three decades. The increase in this value was estimated at about 231 per cent between 1970-71 and 1980-81, 297 per cent between 1980-81 and 1990-91, 233 per cent between 1990-91 and 2000-01 and about 18 per cent between 2000-01 and 200-04. Not only this, the share of livestock products in total value of agriculture and allied activities was also estimated to be growing steadily from 15 per cent in 1970-71 to 26 per cent in 2003-04.

Figure 1: Value Added by Various Livestock Product Groups in India

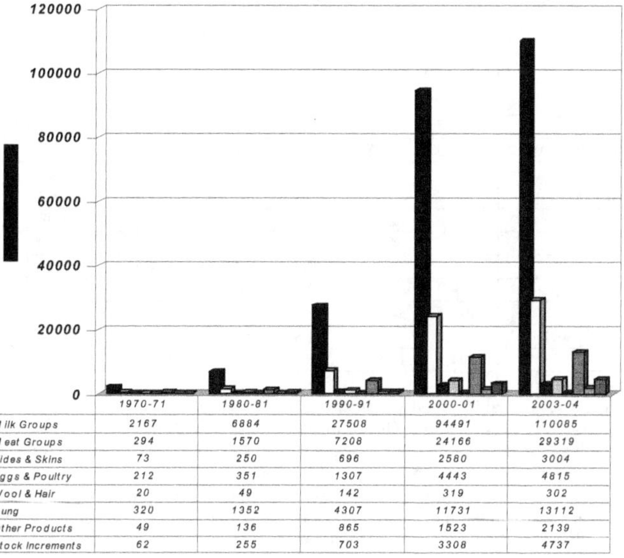

	1970-71	1980-81	1990-91	2000-01	2003-04
■ Milk Groups	2167	6884	27508	94491	110085
□ Meat Groups	294	1570	7208	24166	29319
■ Hides & Skins	73	250	696	2580	3004
□ Eggs & Poultry	212	351	1307	4443	4815
■ Wool & Hair	20	49	142	319	302
▣ Dung	320	1352	4307	11731	13112
▣ Other Products	49	136	865	1523	2139
■ Stock Increments	62	255	703	3308	4737

Year

15

Even at constant prices, the value of livestock products in relation to total value of agriculture and allied activities has grown up over time. The constant price estimates show the share of livestock products in total value of agriculture and allied activities to grow from 19 per cent in 1980-81 to 28 per cent in 2003-04 (Appendix 2). Both at current and constant prices, the increase in value of agricultural products produced in the country has been slower than increase in value of livestock products, particularly between 1970-71 and 2000-01. This is an indication of growing importance of livestock sector in overall agricultural development in the country.

Livestock Population

During the last four to five decades, the livestock economy of India witnessed a number of changes in terms of its size, composition and productivity. The size of livestock herd increased from 307 million in 1956 to 445 million in 1987 and to 485 million in 2003, indicating a significant slowing down in growth of livestock population between 1987 and 2003 as against 1956 and 1987. The slowing down in growth of bovine population between 1987 and 2003 as against 1956 and 1987 has been much sharper compared to total livestock population. The total bovine population in India increased from 204 million in 1956 to 276 million in 1987, and further to only 283 million in 2003, showing, thereby only 0.16 per cent annual increase in the same between 1987 and 2003 as against 0.98 per cent rise between 1956 and 1987 (Table 4).

Table 4: Changing Trends in Livestock population in India: 1956-2003

(million no.)

Species	1956	1966	1987	1997	2003	ACGR@ (%)		
						1956-1987	1987-2003	1956-2003
Total cattle	158.7	176.2	199.7	198.9	185.2	0.74	-0.47	0.33
- Adult female cattle	47.3	51.8	62.1	64.4	-	0.88	-	-
Total buffaloes	44.9	53.0	76.0	89.9	97.9	1.71	1.60	1.67
- Adult female buffaloes	21.7	25.4	39.1	46.8	-	1.92	-	-
Total bovines	203.6	229.2	275.8	288.8	283.1	0.98	0.16	0.70
Sheep	39.3	42.4	45.7	57.5	61.5	0.49	1.87	0.96
Goat	55.4	64.6	110.2	122.7	124.4	2.24	0.76	1.74
Others	8.2	7.5	13.6	16.2	16.0	1.65	1.02	1.43
Total livestock	306.5	343.7	445.3	485.4	485.0	1.21	0.54	0.98
Bovines in livestock (%)	66.4	66.7	61.9	59.5	58.4	-0.23	-0.36	-0.27
Poultry	94.8	115.4	275.3	347.6	489.0	3.50	3.66	3.55
Livestock density@@	2.3	2.5	3.3	3.4	-	-	-	-

Source: (1) Basic Animal Husbandry Statistics, 2002, Department of Animal Husbandry and Dairying, Ministry of Agriculture, Government of India, New Delhi
(2) Government of India, 2005, Annual Report 2004-05, Department of Animal Husbandry and Dairying, Ministry of Agriculture, New Delhi
Note: @-Annual compound growth rates were calculated using the formula: ACGR = $[(P_n / P_0)^{1/n} - 1]$ x 100
where, P_n = Population in current period, P_0 = Population in base period, n = Number of years
@@- Livestock density in per hectare net sown area

Though cattle population in India constituted about 70 per cent of the total bovine stock, the rise in buffalo population over time was much higher. While cattle population grew only by 0.74 per cent a year between 1956 and 1987 with a decline in the same to the extent of 0.47 per cent a year between 1987 and 2003, the increase in buffalo population was 1.71 per cent a year between 1956 and 1987 and 1.60 per cent a year between 1987 and 2000.

In fact, the share of cattle in total bovine population declined from 78 per cent in 1956 to 69 per cent in 2003. This is an indication of the fact that the adult female bovine population has shown considerable rise. Consequently, the sex ratio of adult bovine has shifted rapidly in favour of females. Interestingly, the share of adult female cattle in total cattle population grew from 30 per cent in 1956 to 32 per cent in 1997. On the other hand, adult female buffaloes in total buffalo population showed an increase in share from 48 per cent to 52 per cent between 1956 and 1997. This was perhaps due to increasing farm mechanization in the country. The trend relating to rise in stock of female bovine population is desirable in the light of the growing economic opportunity for increasing milk production and for undertaking dairying as a commercial proposition.

Livestock Production

It is needless to mention that India's livestock sector needs to be encouraged, but not at the cost of cereal production but of course by truly utilizing and exploiting the complementary, supplementary, synergistic and even symbiotic relationship of raising animals with crop production. Encompassing a wide geographical area and reflecting different political system, differing levels of economic development, social systems and changes in tastes, preferences and traditions, the approach to livestock development has varied widely from region to region in India, especially with respect to consumption of milk and milk products. Viewing our dairy spectrum in the light of these variabilities, it becomes pertinent to ask whether the future of our co-operative will remain as bright as in the past if we were to follow the principles and practices of the past.

There is no iota of doubt that since the inception of Operation Flood programme the total milk production has been increasing in all the states of the country. The credit for this healthy scenario should also go to various development projects, which have been simultaneously undertaken to give a fillip to agriculture and dairy production. Though milk production in India has grown significantly over the last two decades, the period gone by is also marked with slowing down in growth of milk production, particularly

17

during the period between 1991-92 and 2003-04 as against the period between 1980-81 and 1990-91 (Table 5). Similarly, there has also been slowing down in growth of egg production during the latter as against the former period. Nonetheless, the growth in egg production in India is much faster as compared to milk production. The growth in wool production in India has remained by and large constant over the past two decades.

Table 5: Changing Trends in Livestock Production in India

Product	TE 1982-83	TE 1992-93	TE 2003-04	ACGR (%)		
				1980-81 to 1990-91	1991-92 to 2003-04	1980-81 to 2003-04
Milk (million tonnes)	33.90	55.87	86.23	5.29	4.02	4.46
- per capita availability (gm./day)	134.33	178.67	228.67	3.10	2.28	2.46
Eggs (billion nos.)	10.80	22.00	41.32	8.80	5.82	6.08
Wool (million kgs.)	33.20	40.53	52.13	2.72	2.69	1.85

Source: Estimates are based on figures compiled from 'Government of India, 2005, Annual Report 2004-05, Department of Animal Husbandry and Dairying, Ministry of Agriculture, New Delhi'
@ - all growth rates are significant at 1 per cent level of probability

A critical evaluation of rates of growth in various livestock products reveals significantly high growth in milk production after 1973-74 with period between 1980-81 and 1990-91 exhibiting the highest growth in milk (5.48 per cent a year) production in India (Table 6). Among various plan periods, the 6th plan shows the highest growth (6.42 per cent a year) in milk production of India. This is an indication of the fact that milk production in India substantially increased between 1980-81 and 1984-85 and thereafter a slowing down in growth of milk production took place. As for the egg production, 6th, 7th and 9th plans registered highest rate of growth. The growth in wool production was significantly high during 6th plan and in the 9th plan. Interestingly, during all the five-year plans beginning 5th plan, the growth in egg production was higher than milk production. The increasing demand for eggs could be one of the reasons for higher growth in production of eggs in India. Further, though annual milk production in India has grown sharply from 34 million tonnes during TE 1982-83 to 56 million tonnes during TE 1992-93, and further to 86 million tonnes during TE 2003-04, the per capita per daily availability of milk estimated at 229 gm during TE 2003-04 is still much lower than the minimum prescribed requirement of 280 gm as recommended by the Indian Council of Medical Research (Figure 1 and Table 5).

Undoubtedly, the programme initiatives undertaken by NDDB and various development projects have led to rise in productivity levels of animals over time. Nevertheless, still there are wide variations in terms of number of bovines (in-milk) and productivity of milch animals across states. And, as a result of this, the milk production is varying considerably across different states of India.

Table 6: Annual Growth Rates of Production of Major Livestock Products – All India

(per cent)

Period	Milk	Eggs	Wool
During pre-and post 'Operation Flood' periods			
1950-51 to 1960-61	1.64	4.63	0.38
1960-61 to 1973-74	1.15	7.91	0.34
1973-74 to 1980-81	4.51	3.79	0.77
1980-81 to 1990-91	5.48	7.69	2.32
1990-91 to 2000-01	4.11	5.67	1.62
During five year plan periods			
5th Plan (1975-76 to 1979-80)	2.91	3.5	1.49
6th Plan (1980-81 to 1984-85)	6.42	8.40	2.67
7th Plan (1985-86 to 1989-90)	4.37	7.23	1.88
8th Plan (1992-93 to 1996-97)	4.41	4.58	0.80
9th Plan (1997-98 to 2001-02)	4.08	7.09	2.20

Source: Basic Animal Husbandry Statistics, 2002, Department of Animal Husbandry and Dairying, Ministry of Agriculture, Government of India, New Delhi

Figure 2: Production and Per Capita Availability of Milk in India

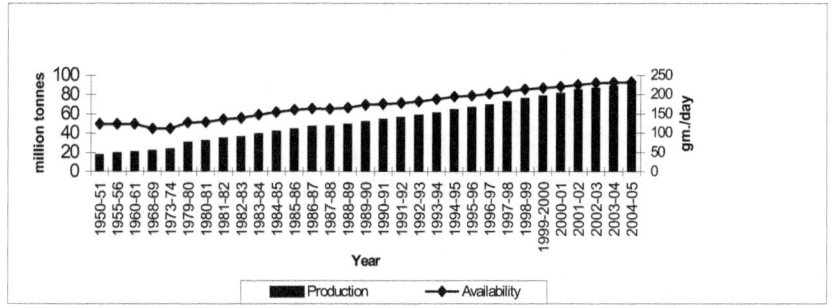

Among various states, though U.P. ranks first in terms of total milk production in India, her share is seen to have stagnated at around 18 per cent over the past two decades (Table 7). A similar pattern is seen in the case of Gujarat, Karnataka, Kerala, Punjab, Tamil Nadu and West Bengal with marginal ups and downs in their share in total milk production of India between 1980-81 and 2003-04. Contrary to this, Maharashtra and Andhra Pradesh have shown a growing trend in terms of their share in total milk production of India, particularly after 1990-91. The states like Madhya Pradesh and Rajasthan in general and Bihar in particular have shown a fall in their share in total milk production of India between 1980-81 and 2003-04. Bihar has shown a steep fall in her share in total milk production of India from 6 per cent in 1980-81 to as low as 3 per cent in 2000-01 with a marginal increase in the same after the 2000-01 period. Changes in

19

Shares of major states in total milk production of India between TE 1982-83 and TE 2003-04 are shown in Figure 3.

Figure 3: Shares of Major States in Total Milk Production of India: TE 1982-83 – TE 2003-04

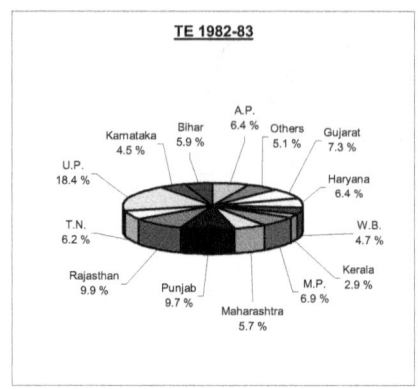 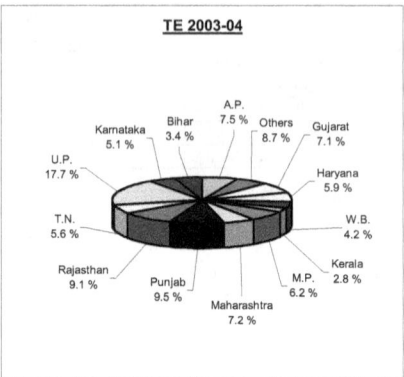

Table 7: Share of States in Total Milk Production of India

(per cent)

State	1980-81	1985-86	1990-91	1995-96	2000-01	2001-02	2002-03	2003-04
Andhra Pradesh	6.4	6.4	5.3	6.4	6.8	6.9	7.6	7.9
Bihar	6.1	5.6	5.1	5.0	3.1	3.2	3.3	3.6
Gujarat	6.8	7.8	6.2	7.0	6.6	6.9	7.1	7.3
Haryana	6.9	5.9	6.3	6.1	6.0	5.9	5.9	5.9
Karnataka	4.5	4.4	4.4	4.8	5.7	5.7	5.3	4.4
Kerala	2.9	2.9	3.0	3.3	3.2	3.2	2.8	2.4
Madhya Pradesh	7.2	6.7	8.0	7.7	5.9	6.3	6.2	6.1
Maharashtra	5.6	5.7	6.8	7.5	7.3	7.2	7.2	7.2
Punjab	10.2	9.3	9.6	9.7	9.6	9.4	9.5	9.5
Rajasthan	10.3	9.4	7.3	8.2	9.2	9.2	9.0	9.1
Tamil Nadu	5.5	6.9	6.0	5.7	6.1	5.9	5.4	5.4
Uttar Pradesh	18.1	18.6	17.3	17.9	17.2	17.4	17.7	18.1
West Bengal	4.1	5.3	5.4	5.0	4.3	4.2	4.2	4.2
Others	5.4	5.1	9.3	5.7	9.0	8.6	8.8	8.9
Total Milk Production (million tonnes)	31.6	44.0	53.9	66.2	80.6	84.4	86.2	88.1

Source: Shares are computed from 'Basic Animal Husbandry Statistics, 1999, Department of Animal Husbandry and Dairying, Ministry of Agriculture, Government of India', and Government of India (2005), Annual Report 2004-05, Department of Animal Husbandry and Dairying, Ministry of Agriculture, New Delhi

Interestingly, while in 1980-81 Bihar ranked higher than Maharashtra in terms of her share in total milk production of India, her share in this respect was found to be one among the lowest in 2003-04. In 2000-01, Bihar showed the lowest share in total milk production of India. On the other hand, in due course of time Maharashtra has overtaken several states in terms of total milk production.

20

Thus, a critical evaluation of Table 7 shows differing scenarios in terms of their contribution to total milk production of India between 1980-81 and 2003-04. In fact, poor growth in breedable bovine population coupled with lower productivity of bovines has greatly affected the milk output growth of several states in the country (Shah, 2002). The cumulative effect of these factors has led to downward slide in their share in total milk production of India over time. A time-scale deceleration in milk production growth caught up with most of the states during the period between 1991-92 and 2003-04 has led to overall deceleration in growth of milk production in India during this period as against the period between 1980-81 and 1990-91. The slowing down in milk production growth of majority of the states in more recent times has raised questions about our ability to meet the domestic requirements in the near future. It becomes, therefore, essential to assess future demand-supply position of the country in milk production.

Demand-Supply Perspective

Several research workers have provided various estimates of India's demand for milk (Kumar and Mathur, 1996; Patel, 1993; Dastagiri, 2003). Among various studies, the study conducted by Dastagiri (2003) relates to estimation of production, consumption and surplus of milk in India for the years 2000 and 2020. While estimating demand, this study assumed 1.63 per cent, 1.54 per cent, 1.40 per cent and 1.51 per cent per annum growth in population during 1993-2000, 2000-10, 2010-20 and 1993-2020, respectively, 1.46 per cent, 3.62 per cent and 3.49 per cent per annum growth in per capita income for rural, urban and pooled, respectively, and pace of urbanization consistent with the recent historical trend. The estimates provided by Dastagiri (2003) with respect to production (supply) and consumption (demand) of milk and various other livestock products are brought out in Table 8.

Although there seems to be an indication of likelihood of generation of significant surpluses of livestock products in the near future, the past trend in terms of employment in livestock sector is not very encouraging (Appendix 3). The past trend between 1983 and 2000 not only shows steady fall in numerical strength of workers engaged in various livestock activities but also in terms of falling share of workers engaged in livestock activities in total workers of the country, both at all-India level and in rural India. This is certainly a discouraging trend that needs to be rectified to strengthen and safeguard this important sector of Indian economy. Though in general employment in livestock sector in India is seen to have declined during 1980s and 1990s periods, this decline may be in

21

certain livestock activities other than milk production as most of the activities in milk business are handled or governed largely by family labour. The dairy sector will therefore remain a focal point for the overall sectoral development of agriculture in India. The efforts made in this sector are paying proposition or paid rich dividends so far. And, India, of late, is witnessing an all pervasive white revolution that can go a long way in ameliorating the lot of those farmers who have positive leaning towards emerging innovative technologies and scientific farm management techniques and also in creating conditions for elimination of regional socio-economic disparities and imbalances.

The advantage in the emerging new technology is that, when optimally utilized, they can truly transform the dairy sector by stepping up milk production to desired levels. The government should, therefore, endeavour to evolve, initiate and implement a more egalitarian policy, especially policies related to procurement pricing and input delivery system that is capable of boosting our livestock production base with all the expediency it deserves.

Table 8: Projections of Livestock Products Production and Domestic Consumption

Year/Livestock Product		2000	2020	Growth rate
Milk	Production	78.56	232.09	5.56
	Consumption	60.77	147.21	4.77
	Surplus	17.79	84.88	
Mutton and goat meat	Production	0.67	9.85	14.40
	Consumption	1.36	12.72	13.62
	Surplus	-1.31	-3.13	
Beef and buffalo meat	Production	3.29	9.11	5.22
	Consumption	0.61	1.15	3.39
	Surplus	2.68	7.96	
Chicken	Production	0.65	2.70	7.38
	Consumption	0.33	0.81	4.72
	Surplus	0.32	1.89	
Egg	Production	32.75	102.91	5.89
	Consumption	13.88	44.06	6.12
	Surplus	8.87	68.85	
Fish	Production	5.66	13	4.40
	Consumption	4.45	8.52	3.30
	Surplus	1.21	4.48	
Pork	Production	0.55	8.22	14.44
	Consumption	N.E.	N.E.	N.E.
	Surplus			
Mutton	Production	0.55	8.21	14.44
	Consumption	N.E.	N.E.	N.E.
	Surplus			
Beef	Production	1.48	5.61	6.87
	Consumption	N.E.	N.E.	N.E.
	Surplus			

Source: Dastagiri (2003)

Since India is likely to generate significant surpluses of milk, the residual after domestic requirement will, therefore, be available for conversion into value added products for exports provided international prices remain favourable. Under such a veritable situation, cooperatives will have to play a crucial role in diverting farmers' produce in the world market. And, among various states, Maharashtra, that ranks first in terms of number and capacity of milk plants operating under central authority as well as state registered authority, will face newer kind of challenges under changed market conditions. The other major states viz., Uttar Pradesh, Gujarat, Rajasthan, Karnataka, Tamil Nadu and Punjab will also face competition in this respect. The achievements of some of the key components of dairy development in different states under cooperative sector are brought out in Appendix 4, which further confirm Maharashtra and Uttar Pradesh, in particular, to be the leading states in the country not only in terms of organised dairy cooperative societies (DCS) but also in terms of procurement, marketing and processing of milk.

In fact, cooperatives have an edge over other competing sectors due to their organizational structure. The federal structure despite many weaknesses provides a very wide network to link many producers to the tertiary level of the economy. Their relevance will now be felt not only in expanding production but also in export trade of milk and milk products. However, how best India can meet the challenges arising in the WTO regime would depend on India's export capabilities and available surplus for exports, aside from favourable international market conditions. In the present milieu, the subsidies extended by European Union countries have created havoc and cheated farmers belonging to developing countries due to depressed international market prices. As long as subsidized production in modern bloc countries continue, the farmers in the developing bloc nations will remain the deprived sections to reap the fruits emanating in the free trade regime due to trade distortions.

Global Trade Distortions or Robbery

One of the important elements of globalisation is the liberalisation of international trade. Increasing flows of livestock and livestock products, including capital, exchange of information, technologies, increasing standards and changes in sectoral structure towards concentration and integration are the major components of globalisation in the livestock sector (FAO, 2005). Of late the distortions in global livestock trade are taking place due to subsidised production of livestock products in EU and USA. These subsidised livestock products are exported in the world markets much below their true cost of production (Sharma et. al. 1996; Williams et. al. 1995, 2004). This coupled with trade barriers, restrictive trade policies and stringent health and sanitary standards restrict many producers in developing world to enter in higher priced international markets (Parthasarathy Rao et. al. 2005).

The USA and EU arrived at a new framework before the fifth WTO Ministerial at Cancun, which could be described as second phase of trade robbery as this framework aimed at further destroying the foundations of food self-sufficiency in some of the developing countries, which is already adversely affected under the consequent impact of the Agreement on Agriculture (AoA). Among various inequalities, the most important one is the manner in which the cattle in the rich countries are over-indulged at the cost of several hundred million farmers in the developing countries. For instance, it has been reported by Sharma (2003) that the EU provides a daily subsidy of US $ 2.7 per cow with Japan providing three times more than US at US $ 8, whereas half of India's 1000 million population live on less than US $ 2 a day. Despite distinct inequalities, the domestic producers in the richest trading block, i.e. Organisation for Economic Cooperation and Development (OECD) were accorded more protective ring in the new agreement. Perhaps Cancun provided this trading block a perfect platform to create further inequalities.

In the dairy sector, the subsidised exports of EU have adversely affected the dairy industry in India, Brazil and Jamaica. For instance, India imported over 1,30,000 tonnes of EU's highly subsidised skimmed milk powder in 1999-2000, which was the outcome of Euro 5 million export subsidies extended by them that works out to approximately 10,000 times the annual income of small-scale milk producer (Sharma, 2003). Further, due to butter export subsidy paid by the EU, butter oil import in India has grown at an average rate of 7.7 per cent annually. This trend has already depressed prices of ghee in the domestic market. Incidentally, despite being the largest producer of milk in the world, India does not provide any subsidy for the dairy sector.

Sharma (2003) further argues that the colourful band of boxes – green box, blue box and amber box are actually tools used by rich countries to protect their subsidies to agriculture with the ultimate goal to dump the surpluses all over the world. It is due to these subsidies that the world markets today are flooded with excess supplies, and consequently resulting in depressed world market prices. Of late, the US has shifted most of the 'blue box' subsidies to 'green box'. Ironically, EU agriculture will continue to be

24

subsidised to the tune of Euro 43 billion for another decade. This amount will further increase with rise in members in WTO folds. In fact, the dairy sector provides livelihood to millions of farmers across India and this sector is most crucial for public health and nutritional requirements. Hence, sustained dumping of cheap imports of dairy products need to be checked by taking stringent measures to revise the tariff rates and quotas.

Notably, in India milk is produced by small farmers belonging to remote areas and processed in plants owned by cooperatives, whereas in EU countries and New Zealand there has been different concept where there stand factory style operations of milk production and squeezing their cows poses a great threat of dumping excess production at lower rates in rest of the world (Gupta, 2001). Since under WTO provisions, there has been greater emphasis on liberalizing trade and government policies to enhance world import demand for dairy products, commitments on market access, reduction of domestic support and subsidies on exports for the removal of distortions in the domestic market, the consequent effect or pressure will be on small farmers, particularly on women (Gupta, 2001). In furtherance, Gupta (2001) indicates that the cost of production of milk in India is one of the lowest in the world while prices of dairy products are among the highest, indicating who has been reaping the fruit so far.

It merits attention here that the WTO was created on January 1, 1995 as a successor to GATT and among various agreements under WTO are the Agreement on Agriculture (AoA) and Agreement on Application of sanitary and phyto-sanitary measures. The major planks of AoA were market access (tarification), domestic support, and export subsidies. According to Parthasarathy Rao et. al. (2005), all Semi-Arid Tropics (SAT) countries are members of WTO and hence are bound by WTO rules under AoA and therefore would be directly influenced by the commitments under AoA by other trading partners.

Although agriculture is characterized by significant levels of government support in many developed countries subject to reduction in commitments under the AoA, these countries continue to support their domestic agriculture due to several loopholes and clauses in the AoA. The range of support is from as high as 60 per cent in many developed countries to negative or below the de minimis level, i.e. the lowest permissible level as specified by WTO. The OECD countries exercise support to their agriculture to the tune of $ 257 billion with the average Producer Support Equivalent (PSE) to the extent of 32 per cent (Table 9). Among various livestock products, the highest PSE was

25

noticed for milk (49 per cent), followed by mutton (42 per cent), beef (35 per cent), pig meat (21 per cent) and poultry meat (17 per cent). Since the chief exporters of dairy and meat products are OECD countries, their high level of protection to these commodities has a large distortionary effect on world trade (Gulati and Narayanan, 2003).

Table 9: Support to Agriculture and Producer Support Equivalents for Livestock Products in OECD Countries: 2003

Commodity	US	EU	OECD
Support to agriculture (US $ million)			
Total agriculture	38878	96549	257285
Livestock	10992	17943	47396
Milk	1197	20389	33598
Beef	367	4736	11032
Pig meat	677	3093	6632
Poultry meat	46	3820	5122
Mutton	166	105	1132
Eggs			
Producer support equivalent (%)			
Total agriculture	18	37	32
Livestock			
Milk	45	51	49
Beef	3	77	35
Pig meat	4	24	21
Poultry meat	4	37	17
Mutton	12	58	42
Eggs	3	2	5

Source: Parthasarathy Rao et. al. (2005)

In fact, as many as 25 developed countries indulged in export subsidies and the EU alone accounted for 88 per cent of the total subsidies, followed by EFTA (European Free Trade Association) countries and the USA (Gulati and Narayanan, 2003). Among various commodities, dairy products, butter oil, cheese and skim milk powder, beef and veal, sugar and coarse grains are subjected to more than 50 per cent of the total export subsidies. Though several countries in South and South East Asia (SSA) and India opted for bound tariff rates (the rates negotiated with WTO prior to joining the organization, which basically represent the upper bound on the level of protection) for crop and livestock products under AoA, there was no reduction commitment as these rates stood higher than the current applied rates. Obviously, the lower rates enabled these countries to raise tariff rates under threat of cheap imports. It is to be noted that domestic support in SAT countries is below the de minimis level and they do not subsidize exports and that a majority of SAT countries in SSA joining the WTO under the least developed country (LDC) are entitled to special privileges, which protect them significantly from global market trade distortions (Parthasarathy Rao et. al. 2005).

It has been argued by Parthasarathy Rao et. al. (2005) that if AoA commitments are implemented without any bias it should lead to reducing support to agriculture in countries with high levels of support, which consequently should lead to curtailing protection in those countries and giving opportunity for exports of production where the support levels are lower or negative. Perhaps this will lead to increase world market prices for primary commodities that include livestock products (Diao et. al. 2001).

As for producer support for livestock extended by developed countries, a similar story unfolds when one goes through the estimates reported by Sharma (2004). During 2001-02, the government support in OECD countries was nearly half of the value of milk production. Ironically, despite countries having lack of comparative advantage in milk production encouraged production through higher support that acted as a limiting factor for consumption as prices stood higher for consumers. The extent of support provided by OECD to their milk producers was US $ 43,393 million during 2001-03 as against US $ 48,107 in 1986-99 (Sharma, 2004). The EU and U.S. also followed similar suit with US $ 18,967 million support provided to milk producers by EU and US $ 11,714 million by US. In fact, various livestock products show significant differences in the level of support and milk is one of the most protected commodities (OECD, 2004). As for milk, while Japan, EU, Canada, Switzerland and US have very high levels of protection; this protection is relatively lower in case of Australia and New Zealand (Table 10). The trends shown by Sharma (2004) are concomitant of the fact that there has hardly been any significant reduction in Producer Support Estimate levels between 1986-88 and 2001-03, indicating limiting impact of WTO on trade distorting support in livestock sector.

Table 10: Producer Support Estimates for Major Livestock Products in Selected OECD Countries

Country	Milk		Beef and veal		Pig meat		Poultry meat		Eggs	
	1986-88	2001-03	1986-88	2001-03	1986-88	2001-03	1986-88	2001-03	1986-88	2001-03
Australia	29	14	6	4	3	3	3	3	37	31
Canada	61	55	10	12	5	7	18	4	22	23
EU	57	47	55	74	16	22	24	37	13	2
Japan	84	77	44	32	42	53	12	11	18	16
New Zealand	9	1	7	1	3	0	56	31	45	26
Switzerland	82	78	78	75	60	67	78	83	80	75
U.S.A.	60	48	6	4	4	4	13	4	9	4
OECD	59	48	32	33	18	21	20	17	18	21

Source: Sharma (2004) and OECD (2004)

In order to dispose off surpluses on to the world markets, export subsidization has turned as the chief policy instrument. Notably, over US $ 27 billion was spent by WTO members for subsidizing exports and out of the 26 countries that committed to reduce

27

export subsidies, EU turned out to be the largest user of export subsidies with 89.4 per cent subsidy expenditure, followed by Switzerland with 5.1 per cent, the US with 1.5 per cent, Norway with 1.3 per cent and other member countries with just 2.7 per cent during 1995-98 (Sharma, 2000). However, export subsidies per unit varied across commodities and countries with a downward trend in export subsidy rates in recent period (Table 11).

Table 11: Export Subsidy Rates in Major Livestock Products in the EU and USA: 1986-90 and 1995-98

Products/Country	1986-90	1995-98	Products/Country	1986-90	1995-98
European Union			*USA*		
Butter and butter oil	248	118	Butter and butter oil	136	58
SMP	106	39	SMP	114	44
Cheese	86	43	Cheese	98	39
Beef and veal	102	57			
Pig meat	111	173			
Poultry	44	20			
Eggs	54	15			

Source: Sharma (2004) and OECD (2004)

Regarding competitiveness of Indian dairy industry 8n the global markets, Sharma and Gulati (2003) clearly show India not being an efficient import substitutes of dairy products as Indian dairy prices compare poorly with the world market prices in terms of Nominal Protection Coefficient (NPC). The reason for this is traced in the nature of world prices of dairy products that have been highly distorted due to export subsidies extended by EU and US. An inter-country comparison of protection levels across major livestock producing countries in the OECD region reveals highest protection being accorded to its milk producers by Switzerland (4.54 in 2001-03), followed by Japan (4.30), Norway (4.25) and 1.82 by OECD region as a whole (Table 12). New Zealand and Australia were not seen to provide any protection to milk producers since the NPC in their case turn out to be unity at the given world market prices.

Table 12: Estimates of NPCs for Major Livestock Products in Selected OECD Countries

Country	Milk		Beef and veal		Pig meat		Poultry		Eggs	
	1986-88	2001-03	1986-88	2001-03	1986-88	2001-03	1986-88	2001-03	1986-88	2001-03
Australia	1.38	1.00	1.00	1.00	1.00	1.00	1.00	1.00	1.18	1.00
Canada	2.61	2.24	1.04	1.02	1.04	1.02	1.19	1.02	1.28	1.28
EU	2.77	1.84	2.25	2.54	1.38	1.25	1.79	1.55	1.24	1.00
Japan	6.49	4.30	1.76	1.44	1.76	2.14	1.14	1.13	1.22	1.19
Korea	3.83	3.05	2.23	3.09	1.50	1.57	2.14	1.58	0.92	1.13
New Zealand	1.02	1.00	1.00	1.00	1.03	1.00	2.83	1.63	1.83	1.35
Norway	3.96	4.25	4.09	5.85	3.77	2.65	2.25	2.97	2.29	1.79
Switzerland	5.51	4.54	4.78	4.08	2.49	3.02	7.28	5.86	6.41	3.69
U.S.A.	2.59	1.85	1.02	1.00	1.00	1.00	1.11	1.00	1.06	1.00
OECD	2.70	1.82	1.41	1.26	1.30	1.22	1.33	1.17	1.22	1.06

Source: Sharma (2004) and OECD (2004)

The foregoing discussion and estimates clearly show that livestock stands one of the highly protected sectors in the developed nations. Though the protection levels have shown a marginally falling trend across nations between 1986-88 and 2001-03, Norway shows an increase in protection level for milk and beef and veal during this period.

As a matter of fact, in 1947-48 a common agreement called as the General Agreement on Tariffs and Trade (GATT) was established by the developed countries to establish a working understanding with respect to export and import of various commodities. Since then there have been eight rounds of talks to modify the agreement. A permanent trade organisation known as WTO was established in 1995 with representation from 124 countries to facilitate fair trade practices among the member countries. And, since India was a signatory of GATT, her membership automatically got ensured in WTO. In fact, there is no harm in India becoming a member of WTO as in the absence of the WTO membership, we could have to negotiate separately with each and every country and that might have placed us at disadvantageous position (Hegde, 2001). Nonetheless, inability to put forward demands strongly during WTO negotiations worked against India. It is to be recollected that during the negotiations in 1985, India failed to bargain and agreed to allow import of milk and milk products under zero per cent bound duty, and due to our new exim policy that brought some essential milk products under Open General License further facilitated import (Hegde, 2001). Though initially there stood no threat as world prices for milk and milk products were ruling high, in due course of time, many developed nations exerted pressure on their governments to provide subsidy to dairy farmers that helped them to reduce price of their dairy products and consequently influenced the world market prices. This provided ample opportunities to the traders to import cheaper milk products with the goal of earning higher profit margins, perhaps at the cost of Indian dairy farmers. This is the plight of Indian dairy farmers in the WTO regime and as long as we continue to remain a member of WTO the threat will continue to bother us. There is hardly any time for Indian dairy farmers to face the challenges of imported milk and milk products under WTO regime and our farmers are yet to be fully geared up to face the situation arising in the era of WTO regime.

World Price Trends Vs Net Social Welfare

A very recent study conducted by Jha (2004) has come out with several interesting observations insofar as the impact of trade liberalization on domestic producers and consumers of milk and milk products is concerned. The study specifically focuses on the fact that though trade liberalization encouraged exports during the 1990s, imports too received impetus following removal of import restrictions in the selected products during mid-1990s. Tariff reduction on milk products, in particular, may have significant implications for the livestock sector since the decision to rear cattle in India is more dependent on the profitability of milk rather than meat.

In order to evaluate the effect of trade liberalization on domestic market of milk and milk products, the study considers three alternative scenarios of world prices (fob) of milk with Rs.647 (US $1500) per quintal as marked with lower range of world price, Rs.1140 (US $ 2650) per quintal as higher range of world price, and Rs.884 (US $ 2050) per quintal as intermediate range of world price.

The study observes a steep fall in the domestic prices of milk following the decision to import milk products at a low international price (Rs.640 per quintal). Since supply of milk is highly price elastic, this situation will lead to adverse affect on milk production in states like Haryana, Maharashtra, Tamil Nadu, West Bengal and Uttar Pradesh that account for more than 40 per cent share in total milk production of India. The decline in milk supply will further translate into negative producer's surplus in all the states. However, the magnitude of negative surplus is likely to vary across states. In this sequel, Uttar Pradesh will show the highest negative surplus, followed by Maharashtra, Tamil Nadu, Haryana and West Bengal (Table 13).

Table 13: Impact of Free Import of Milk on Producer, Consumer and Net Social Welfare in Selected States of India with Different Range of World Prices in 1999

Particulars	Low range of world price (Rs.640 per quintal)					Intermediate (Rs.850 / quintal)	
	Haryana	Maharashtra	T.N.	W.B.	U.P.	Maharashtra	Other States
Production in million quintals	46.8	57.1	45.7	34.7	141.5	57.1	
Producer price – existing	1030	1090	995	1020	960	1090	
Producer price – free trade	824	750	734	759	788	1073	
Supply elasticity with price	0.68	0.74	0.61	0.52	0.58	0.74	
Supply – existing	46.8	57.1	45.7	34.7	141.5	57.1	
Supply under free trade	40.4	43.9	38.4	29.5	126.8	56.5	No change
Change in producers surplus	-8985	-17155	-10981	-8388	-23019	-965.10	
Unit change in producers surplus	-213	-300	-240	-231	-166	-16.9	
Aggregate demand	42.1	57.1	45.7	36.3	138.7	57.1	
Wholesale price – existing	1225	1300	1200	1250	1175	1300	
Wholesale price under free trade	980	895	885	930	965	1280	
Price elasticity of demand	-0.45	-0.47	-0.44	-0.36	-0.36	-0.47	
Existing demand	42.1	57.1	45.7	36.3	138.7	57.1	
Demand under free trade	45.8	65.4	50.9	39.6	147.6	57.4	No chang
Change in consumer's surplus	10779	24819	15227	12151	30064	1146	
Unit change in consumer's surplus	256	435	333	335	217	20.1	
Change in total surplus/welfare	1793	7663	4246	3763	7045	181	
Employment (change in million man-days)	-35.7	-76.2	-42.1	-26.5	-84.8	-3.5	
Forex (change in million US $)	-56.8	-125.4	-79.2	-50.2	-133.9	-8.3	

Source: Jha (2004).

30

The study further finds that at the intermediate range of world price (Rs.850 per quintal), the derived duty paid (DDP) price at one of the Indian port will be Rs.1204 per quintal. On the other hand, price differences in the wholesale market suggest that import would take place only in Mumbai, which would have limited effect on the producers and consumers of milk. The producers and consumers in other states would by and large remain unaffected. At the high range of world price (Rs.1140 per quintal fob at US port), since the ex-port price is significantly higher than the domestic prices, chances of imports at such a high price are less, and, as a result, milk production and consumption will, therefore, be largely unaffected in the country.

The analysis incorporated in this study in general presents a trade-off between the producer's and consumer's interest and establishes that the imports will generally benefit the consumers. However, the losses to the producers also need to be understood adequately. Though the analysis shows that consequent to imports the total economic welfare would increase since a decrease in producer's surplus is over-compensated by the increase in consumer's surplus, the decrease in producers' surplus stands under-estimated in the sense that it ignores loss in employment because of decline in milk production.

The study shows concern for the protection of livestock sector in India in view of loss of employment and the wide ramifications this has for the rural economy. It emphasizes upon the fact that a high import at the low range of the world price would cause enormous loss of employment in the country and, therefore, on this account the sector requires protection from low world price of milk. Since as of now protection in the form of moderate tariff (35-40 per cent) and tariff rate quota appears to be sufficient, any argument for further reduction of tariff must be resisted. The Harbinson's draft in the on-going negotiations on agriculture also recognizes the need for protecting a strategic sector/ sub-sector in a developing country.

Domestic Price Trends

In the new trade regime, pricing issue is of primary importance because on it depends the availability of basic raw material, that is milk. Co-operatives are known for differential seasonal pricing. Thus, while a lean season produce commands a higher procurement price, in flush season the producers are paid relatively less for their produce. This is helpful to both farmers and co-operatives in terms of averaging out of the realization as well as the costs. This type of differential pricing helps in warding off unscrupulous competitors who generally seek to offer a higher procurement price for milk during summers. Though the proportion of milk production during flush and lean seasons varies from 100 and 40, the overall demand remains more or less constant throughout the year. To handle competition, higher price is usually paid for milk during the lean season to attract more supply. This is in tune with the fact that feeds and fodder are in short supply during lean season and, thus, costlier to use, making milk production that much dearer. The producer farmers are, therefore, justified in seeking higher price for milk during lean season to offset increasing cost of production and, in maintaining uniformity in income from milk sales throughout the year even when there may be slump in demand for milk during flush season owing to restricted handling capacity of dairy processing plants and, hence, a lower procurement price on offer for them (Shah, 1996). Further, in due course of time, the planners have favoured a two axis pricing policy for milk that reckons both fat and solids-not fat (SNF) content of milk for its pricing in order to stimulate and give fresh fillip to cow milk production. Though a simple and economically viable technique for SNF testing at village level has been evolved, such a cost effective simple method is not readily available or adopted in or by a significant number of villages of India as of now.

Notably, in the aftermath of liberalisation, conversion of surplus milk to powders helped development of milk production activity in India. Purchase of surplus milk during flush season and conversion to milk powder for recombination during lean season not only formed a major milestone of our development strategy but also helped in expanding market and maintaining a pull on production. In the face of fast changing domestic and international market conditions, taste, preferences and availability of products in processed form, etc., emphasis of late is on meeting the growing market demand for milk powders, butter, ghee, cheese, and other dairy products. Consequent upon rise in demand for products in processed form, the wholesale prices of dairy products have grown faster than wholesale prices of milk and all-commodities (Table 14).

With 1981-82 as the base year, majority of milk and milk products listed in Table 14 showed higher wholesale price index (WPI) as compared to WPI of all-commodities during the entire period between 1990-91 and 1999-2000. Interestingly, while fluid milk, tinned as well as skimmed milk powder, butter and ghee produced in India showed higher growth in their WPI during the first half of the 1990s, this growth in WPI was faster for baby food during the second half of the 1990s period.

Table 14: Wholesale Price Indices of Dairy Products, Food Articles and All-Commodities of India (base 1981-82 = 100)

Product	1990-91	1992-93	1995-96	1997-98	1999-2000	ACGR (%)		
						1991-95	1996-2000	1991-2000
Milk	209.2	264.8	313.8	348.6	403.2	9.86	6.90	6.86
Tinned MP	203.6	305.2	322.2	365.9	442.7	12.00	8.31	7.83
Skimmed MP	178.6	289.9	344.1	372.0	385.6	8.96^{NS}	2.12^{NS}	8.02
Baby food	179.5	211.4	262.0	380.2	423.1	4.89	12.85	10.96
Butter	216.7	262.6	371.1	400.6	452.1	8.66	5.40	8.89
Ghee	188.7	239.1	325.2	342.7	400.4	10.88	5.91	8.35
Cattle feed	155.0	195.8	247.5	289.2	335.5	9.40	7.39	8.72
Fodder	224.4	245.4	326.3	417.3	428.1	4.67^{NS}	7.24	8.30
Food Articles	200.6	271.0	335.7	388.0	457.7	11.10	8.13	9.08
All-commodities	182.7	228.7	295.8	329.8	383.1	10.43	5.37	7.87

Source: Computations are based on 'Basic Animal Husbandry Statistics, 2002', Department of Animal Husbandry and Dairying, Ministry of Agriculture, Government of India, New Delhi

Note: @ - all growth rates are significant at 1 per cent level of probability

NS - growth rates not significant at one per cent level of probability

In general, not only growth in WPI with respect to baby food was higher than growth in WPI of all commodities but also all food articles put together produced in India during the period between 1990-91 and 1999-2000. The trends in WPI with respect to inputs such as cattle feed and fodder were in tune with rise in WPI of milk and milk products. The rise in milk production came about through productivity enhancements, which in turn was due to greater expenditure made on feeds and fodder owing to rise in their costs. The growing trends towards crossbreeding and greater significance accorded to high yielding animals could be the reasons for such relative trends with respect to WPI of milk and milk products and feeds and fodder.

Stakeholders' Perception

The entire dairy business can be categorized into three distinct activities, viz., milk production and its transportation to plants, processing for conversion of milk into pasteurized form and various dairy products, and transportation of products to domestic and export markets. The entire process of movement of milk output from the point of production to the point of consumption in raw/fluid and processed form involves several stakeholders with their varied roles in the process. Although major stakeholders in the dairy policy include milk producers, middlemen, formal and informal sector processors (cooperatives, private, government, etc.), wholesalers, retailers, consumers, etc., in this study only the views expressed by selected milk producers and the secretaries of the selected DCS have been recorded and evaluated in terms of impact of trade related practices indulged in WTO regime on Indian dairy farmers.

In order to extract perceptions of milk producers regarding impact of trade related trends obtaining in WTO regime on their milk business, a quick study/survey was conducted in January 2006 in Kolhapur district of Maharashtra of India. Kolhapur is considered as one of the most developed districts of Maharashtra not only in terms of resource endowments, viz., irrigation network, market infrastructure, etc. but also with respect to awareness of farmers regarding various development activities encompassing input and output prices, policies, market intelligence service for the diversion of products, etc. The study is based on views expressed by secretaries of two selected societies and 30 sampled milk producers drawn from Kolhapur district of Maharashtra.[7] Interestingly, though majority of sampled milk producers were quite aware of the market forces, none among them could air their view in terms of positive or negative impact of globalization and liberalization of policies in the WTO regime. Even the secretaries of the selected societies could not visualize any threat to milk business due to import of milk products at low ruling international market prices. When the milk producers as well as the secretaries of the societies were explained the consequences of cheap imports flowing in the Indian markets due to subsidized production in modern bloc countries that depress their true cost of production and enable them to enter in the world market at much lower prices owing to hidden actual cost, the immediate reaction of these stakeholders drawn from Kolhapur district was in favour of receiving similar kind of subsidies for their milk business.

The stakeholders were quite worried on account of the fact that flow of such cheap imports of milk products, particularly milk powder, might adversely affect their milk production as well as marketed quantities of milk as they could clearly visualize lower prices on offer for their much rich milk output. Under such a veritable situation, these stakeholders were unable to see any incentive to market their produce at unremunerative prices owing to flow of cheap imports of milk powder in the country. As pointed out by Jha (2004), this kind of situation may protect consumer's interest but certainly at the cost of producer's interest leading to adverse affect on domestic production of milk owing to lower ruling prices of milk and milk products in the international market. The selected stakeholders were, therefore, either in favour of receiving similar kind of subsidies as practiced in the European countries or complete abolition of such subsidies practiced in these modern bloc countries. It has already been reported by Sharma et al. (2003) that the competition from subsidized imports will pose a great challenge to the cooperative sector, as they will depress the domestic market prices as well as producers' prices. The increased competition in domestic market from cooperatives coupled with subsidized imports will pose a major challenge to even private sector, and their survival in the fast changing market conditions will become another issue to be addressed. Further, highly subsidized production of dairy products have already led to cost push effect in modern trading bloc in the face of cost pull effect in developing world with little scope for exporting dairy products left with developing nations due to highly distorted and very thin world dairy market, particularly those in EU, USA, Japan and Canada.

Conclusions

Despite constraints like rearing of livestock under sub optimal conditions due to low economic status of livestock owners, India has now become the largest producer of milk in the world. The development of Indian dairy sector is an unprecedented success story as it is based on millions of small producers. Government of India is making concerted efforts to raise the per capita availability of milk through increase in productivity of milch animals. In order to achieve this ambitious goal, assistance is being provided to the State Governments for controlling animal diseases, scientific management and upgradation of genetic resources, increasing availability of nutritious feeds and fodder, improving microbiological quality of milk, etc. The microbiological quality of milk is poor owing to inadequate knowledge of clean milk production and lack of post milking chilling facilities at the village level. In the present milieu, when production of dairy products to match international standards has become necessary to compete in international market of milk and milk products, steps need to be initiated to improve quality of Indian milk products with a view to boost export trade of these products in free trade regime and earn valuable foreign exchange as well as provide clean and quality milk to domestic population for their better health. Though India does not want to leave any stone unturned insofar as her presence in international trade of milk and milk products is concerned, the liberalisation of markets within the WTO framework, especially due to the export subsidies indulged in by OECD countries, now seen to be threatening the Indian dairy sector.

The extent of support by European countries to their dairy industry is of the order of 16 billion Euros a year that works out to more than $ 2 per cow per day subsidy. Ironically, half of the world's population lives less than $ 2 a day. The subsidized exports of milk and milk products indulged in by EU countries are dumped onto the world markets, inflicting stringent and distorted trade environment for developing countries. The direct beneficiaries of these subsidies are the dairy processing and trading companies that, in fact, play with the livelihood of poor farmers belonging to these traditional bloc countries. It could be recollected that a few years ago when India had lowered the tariffs on dairy products, milk products from EU poured into Indian markets, destabilizing domestic dairy sector. The Indian Dairy Association (IDA) then persuaded the Union Government to enhance the import duty to protect the domestic dairy sector.

Evidently, there are serious and distinct differences between developed and developing countries over the issues of subsidization of production as well as export trade and of tariff reduction. Interestingly, while the US-EU bloc is in favour of common tariff reduction formula for all developed and developing countries, the China-India bloc extended support by Brazil, Pakistan and 16 other countries such as Cuba, favoured a different tariff reduction formula for developing countries. Obviously, India and China bloc were in favour of differential treatment for developing countries. However, their proposition and agitation was ruled out by US Deputy Trade Representative Peter Allgier, who was against this two tier system of tariff in WTO and asserted that *"we are for common set of rules for developed and developing countries"*. However, in order to acquire support, he also asserted that the countries like us, just to win the support of LDCs, are ready to give free access in various sectors of their markets. This had obviously created great confusion as he was speaking in favour of developing countries at that particular moment. But, little developing countries could know that the intention was little different and the reality favoured modern bloc countries and that the developing bloc were prone to face the music played by modern bloc nations. This was a ploy of modern bloc countries to create a distorted trade practices in developing world.

In the WTO regime, surging imports have not only affected farm incomes but also employment in many developing countries. Consequent upon cheap imports and absence of adequate protection measures, safeguarding income and livelihood of poor farmers have emerged issues that need to be addressed by policy makers. As rightly pointed out by Sharma (2003), *"the developing world have suddenly become the children of a lesser god and they are the neo-poor"*. Further, though producer milk prices in India stand significantly lower than US and Western Europe, dairy product prices are relatively higher than international market prices owing to domestic processing inefficiencies in India (World Bank, 1999). An earlier study conducted by Sharma et al. (1996) assessed the impact of GATT on dairy products and projected the world dairy product prices to rise by about 2.3 per cent in real terms by the year 2002. The study categorically emphasized on the fact that domestic product prices would remain higher than world market prices despite expected surge in world market prices. This necessitates improvement in production and processing efficiency of livestock sector in India to make it competitive with imports.

As for scope for the expansion of Indian dairy industry in new liberalized trade regime is concerned, it has been argued by Sharma and Sharma (2002) that, in general, the Indian dairy sector would be competitive only if the export subsidies on dairy products are abolished. In more relaxed market environment, the real challenge before Indian livestock sector would be in terms of Sanitary and Phytosanitary Measures (SPS), Agreement on Technical Barriers to Trade (TBT) and animal welfare related issues. With a view to meet these requirements - both domestically and in the world markets - modernization of supply chain covering producer and consumer is the need of the hour.

Undoubtedly, India is already price competitive in the world market and when subsidies from competitive producers like USA and EU countries are removed, the situation will make India more price competitive. In case India is not able to capture the world market in the event of removal of subsidies from the modern bloc countries, the other competitors like Australia and New Zealand would capture this market and enter in a big way to flood markets with their dairy products, making us loosing our competitiveness and a great opportunity in the new trade regime.

End Notes

1. In course of time, India has gradually changed her position from stagnant producer during the decades of 1950s and 1960s to the world's largest producer of milk of late. Though the contribution of agriculture and allied sectors to the gross domestic product (GDP) of India declined from 55 per cent in the early 1950s to 39.5 per cent in 1981-82 and further to 23.9 per cent in 2001-02, the livestock remained among the few high-growth sectors in rural India (Sharma, 2004). This could be evident from the fact that this sector accounted for 25.5 per cent of agricultural GDP and 5.6 per cent of total GDP of India in 2001-02. Not only this, the share of livestock in gross value of agricultural output at 1993-94 prices increased from 18.6 per cent in 1971-72 to 35.5 per cent in 2001-02 (CSO, 2003). The dairy sector contributed the largest share in agricultural GDP.

2. For pursuing objectives relating to promotion of dairy activities in non-operation flood areas, the Department of Animal Husbandry and dairying of the Ministry of Agriculture implemented four schemes in the dairy sector during 2004-05, including a new central sector scheme 'Dairy/Poultry Venture Capital Fund'.

Aside from this, the NDDB continues its activities for the overall development of dairy sector in OF areas.

3. The main objectives of the scheme encompass: (a) development of milch cattle, (b) increase the milk production by providing technical inputs services, (c) procurement, processing and marketing of milk in a cost effective manner, (d) ensure remunerative prices to the milk producers, (e) generate additional employment opportunities, and (f) improve social, nutritional and economic status of residents of comparatively more disadvantaged areas.

4. The new economic policies, ushering in the era of economic liberalization initiated in early nineties, are expected to provide an opportunity to usher in those favourable conditions such as autonomy, freedom with greater accountability and changes in cooperative laws, which will lead to a gradual freeing of the cooperatives.

5. The basic objectives of MMPO are establishing a framework for the orderly disciplined growth of the Indian dairy industry, maintaining the emphasis on supplies of liquid milk rather than manufacture of milk products, ensuring that the intermediaries between the producer and the consumer play a positive and important role towards safeguarding the interests of both, facilitating the operation of the National Milk Grid in balancing the uneven supplies of liquid milk in different regions and seasons through the participation of investor-oriented and privately-owned dairies. The purpose for issuance of the control order is in the interest of the society at large. The promulgation of the MMPO in June 1992 has, therefore, certainly helped many milk Unions since their activities are now confined to collection and distribution of milk.

6. 'Anand Pattern' of dairy co-operatives has a unique position. They are based on six cooperative principles, viz. (1) voluntary membership; (2) democratic decision-making; (3) limited interest on share capital; (4) equitable distribution of surplus; (5) co-operative education and (6) mutual co-operation. The pattern is based on tree tiers of well structured organization with milk producer constituting the smallest unit of the entire business enterprise. The tiers are: the village level dairy co-operative society federating producer members; the district level co-operative milk producers' union federating the village societies and the state level federation constituting of all the district level unions. Of these three tiers, the district dairy co-operative union is the most active unit because it owns the physical infrastructure required for milk procurement, processing and marketing of milk products and for generating inputs required for increasing milk production. The union integrates milk production, procurement and processing and undertakes marketing either independently or with the help of its apex federation.

7. Multistage stratified random sampling technique was used for the selection of *talukas*, villages and milk producing households from Kolhapur district of Maharashtra of India. The *talukas* in the district were classified into two groups as those falling in the eastern and in the western part of the district. Two *talukas*, one of eastern and western parts, were selected purposely from the district. Similarly, based on higher status of DCS in terms of their audit class, literacy level of

farmers, numerical strength of bovines, quantity of milk production, etc., one village from each *taluka* was selected purposely. The village of *Shedshal* belonging to eastern part and *Ghotawade* belonging to western part of Kolhapur district were finally selected for a quick survey. The households in each of the two selected villages, with DCS operating in their village, were categorized as small (1-2 milch bovines), medium (3-4 milch bovines) and large (5 and above milch bovines) based on herd strength using cumulative frequency square root technique (Dalenius and Hodges, 1950) and 15 milk producing households were selected from each village subject to probability proportion to size technique. The selected households were also members of the village milk co-operative society. Out of 30 selected households from two villages, 23 belonged to small category, 4 to medium and the remainig 3 to large category.

References

Bedi, M.S. (1987), 'Dairy Co-operative and Rural Development in India', *Deep and Deep Publications*, New Delhi, India

Central Statistical Organisation (1982), *National Accounts Statistics*, Departments of Statistics, Ministry of Planning, Government of India, New Delhi.

Central Statistical Organisation (1996), *National Accounts Statistics*, Departments of Statistics, Ministry of Planning, Government of India, New Delhi.

Central Statistical Organisation (1997), *National Accounts Statistics*, Departments of Statistics, Ministry of Planning, Government of India, New Delhi.

Central Statistical Organisation (2003), *National Accounts Statistics*, Departments of Statistics, Ministry of Planning, Government of India, New Delhi.

Central Statistical Organisation (2005), *National Accounts Statistics*, Departments of Statistics, Ministry of Planning, Government of India, New Delhi.

Dalenius, T. and Jr. J.C. Hodges (1950), 'The Problem of Optimum Stratification', *Journal of American Statistical Association*, 88-101.

Dastagiri, M.B. (2003), 'Is India Self-Sufficient in Livestock Food Products?', *Indian Journal of Agricultural Economics*, Vol. 58, No. 4, Oct.-Dec., pp. 729-740.

Diao X. A. Somwaru and T. Rao (2001), 'A Global Analysis of Agricultural Reform in WTO Member Countries', pages 25-40 in Agricultural Policy Reform in the WTO-The Road Ahead (Burfisher ME, ed.), *Agricultural Economic Report 802*, United States Department of Agriculture (USDA), Economic Research Service, Beltsville, Maryland, USA.

FAO (2005), 'The Globalizing Livestock Sector: Impact of Changing Markets', 19th Session of the Committee on Agriculture, 13-16 April, 2005, Rome.

Government of India (1999), *Basic Animal Husbandry Statistics*, 1999, Department of Animal Husbandry and Dairying, Ministry of Agriculture, New Delhi.

Government of India (2003), *Basic Animal Husbandry Statistics*, 2002, Department of Animal Husbandry and Dairying, Ministry of Agriculture, New Delhi.

Government of India (2004 a), *Basic Animal Husbandry Statistics*, 2004, Department of Animal Husbandry and Dairying, Ministry of Agriculture, New Delhi.

Government of India (2004 b), *Annual Report 2003-04*, Department of Animal Husbandry and Dairying, Ministry of Agriculture, New Delhi.

Government of India (2005), *Annual Report 2004-05*, Department of Animal Husbandry and Dairying, Ministry of Agriculture, New Delhi.

Gulati, A. and S. Narayanan (2003), 'The Subsidy Syndrome in Indian Agriculture', *Oxford University Press*, New Delhi, India.

Gupta, Shiv Kumar (2001), 'How the Financial Intervention can Reduce the Impact of WTO on Indian Dairy', *Financing Agriculture*, April-June.

Hegde, N.G. (2001), 'WTO Challenges for Indian Dairy Farmers', *Yojana*, December, Vol., 45: 34-35, p. 43.

Jain, M.M. (1979), 'Dairy Development Through Co-operatives: A Study of Rajasthan', *Indian Dairyman*, Vol. 31.

Jha, Brajesh (2004), 'Implications of Trade Liberalization for the Livestock Sector', *Indian Journal of Agricultural Economics*, Vol. 59, No. 3, July-Sept., pp. 566-577.

Kumar, Praduman and V.C. Mathur (1996), 'Agriculture in Future: Demand-Supply Perspective for the Ninth Five-Year Plan', *Economic and Political Weekly*, Vol. 31, No. 39, pp. A131-139.

Kumar, Anjani, Jabir Ali and Harbir Singh (2001), 'Trade in Livestock Products in India: Trends, Performance and Competitiveness', *Indian Journal of Agricultural Economics*, Vol. 56, No. 4, Oct.-Dec., pp. 653-667.

National Dairy Development Board (NDDB), (2001), *Annual Report 2000-01*, Anan, India.

Organization for Economic Cooperation and Development (OECD) (2004), *Agricultural Policies in OECD Countries at a Glance, 2004*, Paris, France.

Parthasarathy, P. Rao, P.S. Birthal and J. Ndjeunga (2005), 'Crop-Livestock Economies in the Semi-Arid Tropics: Facts, Trends and Outlook, *International Crops Research Institute for the Semi-Arid Tropics*, Patancheru, India.

Patel, R.K. (1993), 'Present Status and Promise of Dairying in India', *Indian Journal of Agricultural Economics*, Vol. 48, No. 1.

Shah, Deepak (1996), 'Working of Milk Producers' Co-operatives in Maharashtra', *Mimeograph Series No. 41*, Gokhale Institute of Politics and Economics, Pune.

Shah, Deepak (1997), 'Co-operative Dairying in Maharashtra: Lessons to be Learned', *Economic and Political Weekly*, Vol. 32, No. 39, September 27, pp. A-125-A-135.

Shah, Deepak (2001), 'Indian Dairy Industry: Present Status and Future Prospects', *Productivity,* Vol. 42, No. 3, October - December, pp. 474-483.

Shah, Deepak (2002), 'Milk Production in India: An Analysis of Spatial and Temporal Variations', *The Asian Economic Review*, Vol. 44, No. 2, August, pp. 291-304.

Sharma, R., Konandreas P. and Greenfield J. (1996), 'An Overview of the Assessment of the Impact of the Uruguay Round on Agriculture Prices and Incomes', Food Policy, 21 (4/5), 351-363.

Sharma, Vijay Paul (2000), 'Assessing the Effects of the WTO Agreement on Indian Dairy Industry: What Can We Learn from Past Five Years?', *Indian Dairyman*, Vol. 52, No. 11, November, pp. 7-26.

Sharma, Vijay Paul and Pritee Sharma (2002), "Trade Liberalization and Indian Dairy Industry", *Oxford & IBH Publishing Co. Pvt. Ltd.*, New Delhi.

Sharma, Vijay Paul, Christopher L. Delgado, Steve Staal and Raj Vir Singh (2003), 'Policy, Technical and Environmental Determinants and Implications of the Scaling-up of Milk Production in India', Report submitted as a part of Phase-II IIMA-IFPRI study under the IFPRI-FAO project entitled 'Livestock Industrialization, Trade and Social-Health-Environment Impacts in Developing Countries', funded by the *Department for International Development (DFID)*, U.K., through the Livestock, Environment and Development (LEAD) initiative at FAO.

Sharma, Vijay Paul and Ashok Gulati (2003), 'Trade Liberalisation, Market Reforms and Competitiveness of the Indian Dairy Sector, MTID Discussion Paper No. 61, Markets, Trade and Institutions Division, International Food Policy Research Institute, Washington, D.C., U.S.A.

Sharma, Vijay Paul (2004), 'Livestock Economy of India: Current Status, Emerging Issues and Long-Term Prospects', *Indian Journal of Agricultural Economics*, Vol. 59, No. 3, July-Sept., pp. 512-554.

Sharma, Devinder (2003), 'WTO and Agriculture: The Great Trade Robbery', *http://www.indiatogether.org/2003/sep/dsh-robbery.htm, or http:// www.countercurrents.org/en-Sharma20903.htm*

Williams T.O., J.M. Powell and S. Fernandez-Rivera (1995), 'Manure Utilization, Draught Cycles and Herd Dynamics in the Sahel: Implications for Cropland Productivity', Pages 392-409 in Livestock and Sustainable Nutrient Cycling in Mixed Farming Systems of Sub-Saharan Africa (Powell, J.M., Fernandez-Rivera, S., Williams T.O. and Renard, C., eds.). Volume II: Technical Papers. Proceedings of an International Conference on Livestock and Sustainable Nutrient Cycling in Mixed Farming Systems of

Sub-Saharan Africa, Addis Ababa, Ethiopia, 22-26 Nov. 1993, ILCA, Addis Ababa, Ethiopia.

Williams T.O., S. Ehui and P. Kormawa (2004), 'Implications of Changing Domestic Policies and Globalization for Crop-Livestock Systems Development in West Africa'. Paper presented at the International Conference: Sustainable Crop-Livestock Production for Improved Livelihoods and Natural Resource Management in West Africa, 19-22 Nov., 2001, Ibadan, Nigeria.

World Bank (1999), 'India Livestock Sector Review: Enhancing Growth and Development, Rural Development Sector Unit, South Asia Region', *Allied Publishers Pvt. Ltd.*, New Delhi, India.

Appendix 1: World Estimates of Milk Production: 1990-2003

(million tonnes)

Year	Cow	Buffalo	Goat	Sheep	Total *	India's Share in World
1990	479.2 (88.3)	44.1 (8.1)	10.0 (1.8)	8.0 (1.5)	542.5 (100.0)	53.7 (9.9)
1991	470.0 (88.1)	44.4 (8.3)	9.8 (1.8)	7.9 (1.5)	533.3 (100.0)	54.1 (10.1)
1992	460.7 (87.6)	46.1 (8.8)	10.4 (2.0)	7.8 (1.5)	526.1 (100.0)	56.4 (10.7)
1993	460.2 (87.1)	48.2 (9.1)	10.9 (2.1)	7.7 (1.5)	528.1 (100.0)	58.9 (11.2)
1994	461.3 (86.6)	50.5 (9.5)	11.3 (2.1)	8.0 (1.5)	532.4 (100.0)	61.4 (11.5)
1995	463.8 (86.0)	54.4 (10.1)	11.7 (2.2)	8.0 (1.5)	539.1 (100.0)	64.6 (12.0)
1996	467.4 (85.6)	57.2 (10.5)	11.8 (2.2)	8.3 (1.5)	545.9 (100.0)	67.3 (12.3)
1997	469.9 (85.3)	59.5 (10.8)	12.1 (2.2)	8.2 (1.5)	550.9 (100.0)	71.1 (12.9)
1998	476.9 (85.0)	62.9 (11.2)	11.6 (2.1)	8.2 (1.5)	560.8 (100.0)	75.2 (13.4)
1999	483.4 (84.9)	65.5 (11.5)	11.4 (2.0)	8.2 (1.4)	569.7 (100.0)	78.1 (13.7)
2000	490.6 (84.7)	67.6 (11.7)	11.6 (2.0)	8.0 (1.4)	579.1 (100.0)	81.0 (14.0)
2001	495.8 (84.6)	68.8 (11.7)	11.8 (2.0)	8.1 (1.4)	585.8 (100.0)	82.0 (14.0)
2002	505.7 (84.7)	70.5 (11.8)	11.8 (2.0)	8.0 (1.3)	597.4 (100.0)	84.0 (14.1)
2003	506.9 (84.4)	72.7 (12.1)	12.0 (2.0)	8.1 (1.3)	600.9 (100.0)	87.0 (14.5)
* Milk total for World includes Camel milk						
Figures in brackets show percentages to total production						
Source: FAOSTAT - Website						

Appendix 2: Value Added from Agriculture and Allied Activities in India: Constant Prices

(Rs. crores)

Sr. No.	Items	At 1980-81 Prices			At 1993-94 Prices		
		1980-81	1990-91	1994-95	1993-94	2000-01	2003-04
1.	Value of Output	56875	80565	90671	271839	320920	346538
	1.1 Agriculture	46278	63383	70181	204874	235469	251230
	1.2 Livestock	10597	17182	20490	66965	85451	95308
	Share (%)						
	a. Milk Groups	64.96	68.16	67.45	64.82	67.99	65.92
	b. Meat Groups	14.82	15.54	16.28	18.72	16.63	18.15
	c. Eggs	3.31	4.16	4.24	3.41	3.77	3.80
	d. Wool & hair	0.46	0.37	0.36	0.27	0.26	0.23
	f. Dung	12.76	8.89	7.85	9.28	7.80	7.53
	e. Other Products	1.28	1.58	1.39	1.45	1.18	1.27
	f. Increment in Stock	2.41	1.30	2.43	2.05	2.37	3.10
2.	Less Inputs	15247	20695	23176	55401	64815	69602
3.	Gross Domestic Product	42466	60991	68706	221834	262196	283323
4.	Less Consumption of Fixed Capital	2410	3218	3830	10915	13600	14971
5.	Net Domestic Product	40056	57773	64876	210919	248596	268352
6	Share in Value of Output						
	- Agriculture	81.37	78.67	77.40	75.37	73.37	72.50
	- Livestock	18.63	21.33	22.60	24.63	26.63	27.50
7.	Percent increase of 1.1 over						
	- 1980-81	-	36.96	51.65	-	-	-
	- 1990-91	-	-	10.73	-	-	-
	- 1993-94	-	-	-	-	14.93	22.63
	- 2000-01	-	-	-	-	-	6.69
8.	Percent increase of 1.2 over						
	- 1980-81	-	62.14	93.36	-	-	-
	- 1990-91	-	-	19.25	-	-	-
	- 1993-94	-	-	-	-	27.61	42.33
	- 2000-01	-	-	-	-	-	12.71

Source: Estimates are based on figures Compiled from National Accounts Statistics, Central Statistical
Organisation, Ministry of Planning and Programme Implementation, Government of India, 1997 and 2005

Appendix 3: Sectoral Growth in Employment by Usual Activity Status: 1983-2000

(in '000' nos.)

Sector/Activity	NIC Code 70/87	All India				Rural			
		1983 (38th Round)	1987-88 (43rd Round)	1993-94 (50th Round)	1999-2000 (55th Round)	1983 (38th Round)	1987-88 (43rd Round)	1993-94 (50th Round)	1999-2000 (55th Round)
Agriculture	0	178277 (66.32)	185922 (63.91)	207576 (62.52)	193766 (57.60)	171088 (80.00)	174832 (76.60)	195757 (76.90)	187852 (72.02)
Livestock		11973 (4.45)	12380 (4.26)	9789 (2.95)	8027 (2.40)	10436 (4.88)	10727 (4.70)	7891 (3.10)	7958 (3.05)
Mining	1	1730 (0.64)	2139 (0.74)	2684 (0.81)	2026 (0.60)	1133 (0.53)	1369 (0.60)	1782 (0.70)	1410 (0.54)
Manufacturing	2 & 3	29390 (10.93)	32510 (11.17)	35451 (10.68)	36487 (10.84)	14756 (6.90)	17346 (7.60)	18074 (7.10)	19358 (7.42)
Electricity, Gas, Water	4	850 (0.32)	1032 (0.35)	1312 (0.40)	893 (0.27)	342 (0.16)	456 (0.20)	509 (0.20)	366 (0.14)
Construction	5	6642 (2.47)	11598 (3.99)	11512 (3.47)	15405 (4.58)	3978 (1.86)	8217 (3.60)	6619 (2.60)	9174 (3.52)
Trade	6	17920 (6.67)	21345 (7.34)	26287 (7.92)	34738 (10.32)	7891 (3.69)	10043 (4.40)	11710 (4.60)	14241 (5.46)
Transport	7	7261 (2.70)	8186 (2.81)	10209 (3.08)	12712 (3.78)	2588 (1.21)	3424 (1.50)	4073 (1.60)	5937 (2.28)
Services	8 & 9	25563 (9.51)	28030 (9.63)	36709 (11.06)	32555 (9.67)	11377 (5.32)	12553 (5.50)	15783 (6.20)	14515 (5.57)
Total Employed Workers		268820	290930	332000	336610	213860	228240	254560	260812

Source: http://www.dahd.nic.in/
Notes: Figures in parentheses are percentages to total workers
* NSS 50th round onwards National Industrial Classification 1987 (NIC-87) codes are in use.

Appendix 4: Achievements of Some of the Key Components of Dairy Development in Different States Under Cooperative Sector as on March 31, 2004

Sr. No.	States and UT	DCS Organised (nos.)	Farmer Members ('000')	Rural Milk Procurement ('000' Kg/Day)	Liquid Milk Sale ('000' LD)	Processing Capacity ('000' LD)
1.	Andhra Pradesh	5590	785	953	898	2150
2.	Assam	65	2	3	8	60
3.	Bihar	4621	239	402	447	666
4.	Delhi	-	-	-	1937	1350
5.	Goa	169	19	43	89	75
6.	Gujarat	12112	2580	5052	2107	6720
7.	Haryana	4257	230	326	152	530
8.	Himachal Pradesh	235	17	25	16	40
9.	Karnataka	9311	1741	2243	1517	2530
10.	Kerala	3218	706	614	740	905
11.	Madhya Pradesh	5089	250	313	323	1000
12.	Maharashtra	18349	1583	2683	2648	4650
13.	Nagaland	76	3	2	4	10
14.	Orissa	1654	122	127	132	185
15.	Pondicherry	93	30	55	52	50
16.	Punjab	7283	404	745	496	1545
17.	Rajasthan	9643	534	1035	855	1295
18.	Sikkim	189	7	9	7	15
19.	Tamil Nadu	7578	1988	1664	1206	2601
20.	Tripura	84	4	2	9	10
21.	Uttar Pradesh	17826	778	797	436	1670
22.	West Bengal	2287	172	327	823	1600
	Total	109729	12194	17420	14902	29657

Source: http://www.dahd.nic.in/
Note: DCS = Dairy Cooperative Societies; LD = Litres Per Day

44

Publisher: Eliva Press SRL

Email: info@elivapress.com

www.elivapress.com

www.ingramcontent.com/pod-product-compliance
Lightning Source LLC
Chambersburg PA
CBHW051253170526
45165CB00004B/1689